Grenfell and Construction Industry Reform

In the wake of the tragic events of the fire at Grenfell Tower, the inquiry into the fire and the independent Hackitt Review revealed deep-rooted and unpalatable truths about the current state of the UK construction industry. Dame Judith Hackitt was scathing in her assessment of the construction industry denouncing it as "an industry that has not reflected and learned for itself, nor looked to other sectors" and defining the key issues as ignorance, indifference, lack of clarity on roles and responsibilities and inadequate regulatory oversight and enforcement tools.

There is an urgent need to change practices and behaviours to prevent a similar tragedy from reoccurring. This book sets out the changes required, why they are required, how they are to be achieved and the progress towards them to date.

Implementation of these major safety reforms will move the construction industry from the conditions that allowed the fire at Grenfell Tower to occur, to a system where construction professionals take greater responsibility for the safety of residents in their buildings. This book provides an overview of how the movement towards implementing a new building safety regime has unfolded over the last three years and details what still needs to be done if residents are to feel safe and be safe in their own homes.

Steve Phillips is a Chartered Building Surveyor and Chartered Building Engineer who is a Senior Lecturer in Building Surveying at Northumbria University. He also sits on the editorial board of the Journal of Building Survey, Appraisal & Valuation.

Jim Martin is Executive Chairman of the Martin Arnold Group, UK. He is Chair of the Housing Forum's Working Group: Smarter Supply: Smarter Homes and Conference Panel Member: Implementing the Hackitt Review: Next Steps (2019).

Grenfell and Construction Industry Reform

A Guide for the Construction Professional

Steve Phillips and Jim Martin

Routledge
Taylor & Francis Group

LONDON AND NEW YORK

First published 2022
by Routledge
2 Park Square, Milton Park, Abingdon, Oxon OX14 4RN

and by Routledge
605 Third Avenue, New York, NY 10158

Routledge is an imprint of the Taylor & Francis Group, an informa business

British Library Cataloguing-in-Publication Data
A catalogue record for this book is available from the British Library

Library of Congress Cataloging-in-Publication Data
Names: Phillips, Steve (Chartered Building Engineer), author. |
Martin, Jim (Executive Chairman of the Martin Arnold Group,
UK), author.
Title: Grenfell and construction industry reform : a guide for the
construction professional / Steve Phillips and Jim Martin.
Description: Abingdon, Oxon ; New York, NY : Routledge,
2022. | Includes bibliographical references and index.
Identifiers: LCCN 2021012596 (print) | LCCN 2021012597 (ebook)
Subjects: LCSH: Industrial policy—Great Britain. | Construction
industry—Government policy—Great Britain. | Construction
industry—Safety regulations—Great Britain. | Building—
Standards—Great Britain. | Grenfell Tower fire, London, England,
2017. | Fire investigation—Great Britain.
Classification: LCC HD9715.G72 P55 2022 (print) |
LCC HD9715.G72 (ebook) | DDC 338.4/76240941—dc23
LC record available at https://lccn.loc.gov/2021012596
LC ebook record available at https://lccn.loc.gov/2021012597

ISBN: 978-0-367-55284-8 (hbk)
ISBN: 978-0-367-55285-5 (pbk)
ISBN: 978-1-003-09280-3 (ebk)

DOI: 10.1201/9781003092803

Typeset in Bembo
by codeMantra

Contents

Figures

List of acronyms

ACM	Aluminium Composite Material
ACP	Aluminium Composite Panel
AIM	Asset Information Model
AN	Advice Note
AP	Accountable Person
BIM	Building Information Modelling
BRE	Building Research Establishment
BSA	Building Societies Association
BSM	Building Safety Manager
BS	British Standard
BSR	Building Safety Regulator
BSCC	Building Safety Competence Committee
CAD	Computer Aided Design
CCP	Chartered Construction Professional
CCS	The Considerate Constructors Scheme
CDE	Common Data Environment
CDM	Construction Design & Management
CE	Conformité Européenne
CIA	Chemical Industries Association
CIBSE	The Chartered Institution of Building Services Engineers
CPD	Continuing Professional Development
CSG	Competence Steering Group
DCLG	Department for Communities and Local Governments
DLT	Distributed Ledger Technology
EN	Explanatory Notes
EPDM	Ethylene Propylene Diene Monomer
EWS	External Wall System
FEF	Fire Emergency File
FR	Fire Retardant
FRA	Fire Risk Assessments

HRRB	Higher Risk Residential Building/High Rise Residential Buildings
HSE	Health & Safety Executive
IA	Impact Assessment
IET	Institution of Engineering & Technology
ISO	International Standards Organisation
JCA	Joint Competent Authority
KCTMO	The Kensington and Chelsea Tenant Management Organisation
LABS	Local Authority Building Standards
LFB	London Fire Brigade
LGA	Local Government Association
MHCLG	Ministry of Homes Communities & Local Governments
NFCC	National Fire Chiefs Council
PAS	Public Approved Specification
PE	Polyethylene
PII	Professional Indemnity Insurance
PIM	Project Information Model
PIR	Polyisocyanurate
QR	Quick Response
RBKC	Royal Borough of Kensington and Chelsea
RES	Resident Engagement Strategy
RFID	Radio Frequency Identification
RICS	Royal Institution of Chartered Surveyors
RP	Responsible Person
RRO	Regulatory Reform Order
UK	United Kingdom
uPVC	Unplasticized Polyvinyl Chloride
WG	Working Group

Acknowledgements

I would like to thank Northumbria University for allowing me the time to write this book.

I am extremely grateful for the help provided by Liam Graye MCABE for allowing me to use his research on competency frameworks.

I am extremely appreciative of the assistance provided by Barbara Locke, who very generously gave hours of her time to proofread the draft versions of the book.

Introduction

Following the tragic events of the fire at Grenfell Tower, both the inquiry into the fire and the independent Hackitt Review of the deficiencies of the building regulatory framework for high-risk residential buildings (HRRBs) revealed a number of deep-rooted and unpalatable truths about the current state of the UK construction industry. Dame Judith Hackitt was scathing in her assessment of the construction industry describing it as "*an industry that has not reflected and learned for itself, nor looked to other sectors*". She defined the four key issues that underpinned the systemic failure as ignorance, indifference, lack of clarity on roles and responsibilities and inadequate regulatory oversight and enforcement tools. In Hackitt's view, these four issues combined to create an inherent cultural problem within the construction industry, resulting in what she denounces as a "race to the bottom".

Clearly, as an industry, we need to change our practices and behaviours to prevent a similar tragedy from reoccurring. The main objective of this book is to set out the changes required, why they are required, how they are to be achieved and the progress towards them to date. Implementation of these reforms will move us from the conditions that allowed the fire at Grenfell Tower to occur, to a system where construction professionals take greater responsibility for the safety of residents in their buildings.

Reform on this scale certainly requires new legislation, but it must combine with a significant culture change in the industry and increased levels of competence for all those operating within the sector. The UK construction industry needs to begin the change process immediately. The scale of the change required is so far-reaching and so urgent that the need for this book is also immediate rather than waiting for the draft Building Safety Bill to be enacted or for Phase 2 of the Grenfell Inquiry to be completed.

DOI: 10.1201/9781003092803

This book is not a substitute for reading primary source documents such as the Final Hackitt Review or the draft Building Safety Bill, but rather provides an overview of how the movement towards implementing a new building safety regime has unfolded over the last three years and details what still needs to be done if residents are to feel safe and be safe in their own homes.

The book comprises six chapters which can be summarised as follows:

Chapter 1: The Grenfell Tower fire

The opening chapter examines the refurbishment works and changes made to the original tower at Grenfell, including the installation of a new external cladding facade system to improve the energy efficiency of the building. It explains how the outbreak of a standard fire in a fridge-freezer of one of the residential flats allowed flames and smoke to escape into the external cladding façade system with disastrous consequences.

The chapter also explores the findings of phase one of the Grenfell Inquiry and the evidence presented to date in phase two of the inquiry, exemplifying the overall systemic failure within the UK construction industry that ultimately led to the tragic fire at Grenfell.

Chapter 2: The Hackitt Review

This chapter centres on Dame Judith Hackitt's independent review of building regulations and fire safety. The review has a specific focus on the application of building regulations and fire safety to high-risk residential buildings and the need to provide assurance to residents that the buildings they live in are safe and will remain safe. Hackitt published an interim report in December 2017 which concluded that the current regulatory system for ensuring fire safety in high rise and complex buildings is not fit for purpose and that a radical overhaul was required of the culture of the construction industry and the effectiveness of the regulators.

The final review, which was published on 17 May 2018, is explored in depth including the review's objective to address the systemic failure through a new regulatory framework for the construction and management of high-rise residential multi-occupancy buildings. Criticism of Hackitt's Final Review is also explored as is Dame Hackitt's response to that criticism.

Chapter 3: The Government response and the draft Building Safety Bill

Following the publication of the final Hackitt Review, the Government accepted all its findings and the 53 recommendations of the review. In December 2018, the Government published an implementation plan, setting out how they proposed to take forward these recommendations. The key proposals in the plan set out a stronger, more effective regulatory framework, with defined risk ownership, clearer guidance on building standards and better engagement with residents, giving them a stronger voice. The end goal was to create a paradigm shift in culture across the construction industry. In June 2019, the Government published and advertised the "Building a Safer Future" consultation which set out how the necessary long-term reform could be achieved. The results of this consultation were published in April 2020 and, at the core of the proposals, is a new stringent regulatory regime for higher-risk buildings, which forms the framework for the draft Building Safety Bill. Chapter 3 provides an explanatory commentary on the draft bill and details how the new Building Safety legislation proposes to introduce accountable duty-holders and three 'Gateways' to ensure that building safety risks are considered during the planning, design and construction processes. The requirement to create and store a 'golden thread of information' using a digital platform and maintain it throughout the lifecycle of a building is introduced in this chapter. Finally, Chapter 3 considers how the residents are to be placed at the heart of the -process in order to provide them with a voice that is clearly heard.

Chapter 4: Competency frameworks

This chapter describes the work of the Competence Steering Group (CSG) and the twelve working groups that were formed "*to come up with a blueprint to improve competence for those working on higher risk buildings and drive a culture change right across the industry*". The chapter describes the new regime as defined in the "Setting the Bar" report which puts in place a comprehensive framework of competence standards on an individual sector basis, supported by third-party assessment for both individuals and companies and corresponding third-party accreditation of those who are charged with undertaking the assessments. The chapter also examines the report's proposal for an overarching system of competence which includes a new competence committee sitting within the Building Safety Regulator, a national suite of competence standards and independent assessment against these competence standards. The implications of the final report produced by the CSG's Working Group 8, entitled *Safer People, Safer Homes: Building Safety Management*, are also set out in Chapter 4.

Chapter 5: The golden thread and traceability

The Hackitt Review recommends a very clear model of risk ownership, with transparency of information and an audit trail available all the way through the life cycle of a building, to provide reassurance and evidence that a building was built safely and continues to be safe in occupation. An important component of this audit trail will be the creation of a digital record to create "a golden thread of information" specific to each higher-risk building. The Hackitt Review also highlights confusion over product labelling as a contributory factor to fire safety systems being compromised and recommends that the digital record should also be used to provide product traceability. Chapter 5 explores the different types of digital technology that can be used to support both the "golden thread of information" and the traceability of building products.

Chapter 6: Four years on

The final chapter tracks the practical progress that has been made towards ensuring residents of HRRBs feel safe and are safe in their homes. The chapter opens with a review of the independent expert advisory panel set up to provide advice and make recommendations to the Secretary of State for Housing, Communities and Local Government on urgent building safety measures that should be carried out to HRRBs with Aluminium Composite Material external wall cladding. The chapter covers the unintended consequences of the panel's much-criticised "Advice Note 14" for the valuation of flats in high-risk residential buildings, as many valuation surveyors take the position that if compliance with the advice document cannot be demonstrated, then all flats within that building will have a valuation of £nil. The industry's response was to produce the EWS1 form which was intended to unblock the market. However, this aim has not been achieved, and the final chapter details why the EWS1 form is failing and sets out alternative ways of restoring confidence within the market.

1 The Grenfell Tower fire

Background

Grenfell Tower is approximately 67 m in height and has 25 storeys including a basement. The building comprises an in situ concrete structural frame with a central reinforced concrete core, reinforced concrete floors and perimeter reinforced concrete columns. Originally, pre-cast concrete spandrel panels formed the cladding to the upper 20 storeys of the building, with sliding windows units of mill-finished, single glazed aluminium and non-structural white window infill panels. At the top of the building was a pre-cast architectural crown with tapered pilasters at the tops of the columns and a ring of freestanding concrete beams. In the central core of the building was a single staircase and two lifts serving each floor of the tower. The residential flats occupied floors 4–23 with 6 flats on each floor. The flats were separated with reinforced concrete cross walls. The lower levels of the building were designed for use by the local community.[1] Figure 1 shows the external view of Grenfell Tower prior to the refurbishment works being carried out.

The refurbishment

Grenfell Tower is owned by the Royal Borough of Kensington and Chelsea (RBKC), but from 2009, the management of the building became the responsibility of an independent company, The Kensington and Chelsea Tenant Management Organisation (KCTMO). Between 2012 and 2016, an extensive regeneration project altered the building both internally and externally. These works included the refurbishment of the lower part of the building and the creation of nine new flats. Building services work was undertaken throughout the building including the installation of a new heating system and the modification

DOI: 10.1201/9781003092803-1

Figure 1 Grenfell Tower Prior to the Refurbishment Works.
(Credit: Dr Lane supplemental report [BLAS0000008] p. 6 Fig. 8.2)

of the original smoke control system to become a new combined environmental and smoke control system. The most significant works were the over- cladding of the existing building with a new insulation and rainscreen cladding system.

The use of over-cladding on tower blocks was a standard solution used by Local Authorities at the time to meet their obligations under the Decent Homes Programme[2] to ensure that all homes within their stock should be in a good state of repair and have effective heating and insulation. The design and access statement in the planning application for the works at Grenfell Tower stated that the changes to the envelope of the building were part of an integrated solution to tackle the building's inefficient energy performance. The main objectives for the proposed over-cladding system were set out as:

- A dramatic improvement in heat loss with new insulation and air sealing which will generate significant energy savings
- New windows units which can naturally vent the building throughout the year, provide natural daylighting, and be safely cleaned from the inside
- Improvement in the appearance of the tower

Originally, Leadbitter Construction were the preferred contractor to carry out the proposed works, but at £11.278 million (inclusive of fees), their bid was £1.6 million above the proposed budget. After a further procurement process in June 2014, Rydon Maintenance Limited were appointed as the design and build contractor with a bid of £8.7 million. They contracted Harley Facades Ltd as their cladding sub-contractor. The external cladding was supplied by Arconic with the two main suppliers of insulation being Kingspan and Celotex. Studio E was the architect for the main refurbishment works, and the Employer's agent/Quantity Surveyor was Artelia Projects UK Limited. A degree of specialist fire engineering services was provided by Exova Warrington Fire Consultants.

The project was funded by RBKC, and the Department of Building Control at RBKC acted as the building control authority. They undertook a number of inspections between August 2014 and 2016, and the building certificate for the completion of works being signed off on 7 July 2016.

In addition to the main refurbishment works, the KCTMO had, between 2011 and 2013, carried out a programme of replacing 106 entrance doors to the flats owned by RBKC tenants with fire doors that complied with the relevant legislative fire safety standards. The manufacture of the doors and the installation works were carried out by Manse Masterdor.

The fire

In the early hours of the morning of 14 June 2017, a fire reported to have started in a fridge-freezer located in Flat 16 on the fourth floor of Grenfell Tower. The fire moved beyond the kitchen as flames and hot gases caused the uPVC jamb of the kitchen window to deform and collapse. This provided an opening for the flames to pass into the cladding façade system on the outside of the building. The cladding consisted of aluminium composite material (ACM) rainscreen panels bonded to a central polyethylene core which, being highly combustible, was the primary cause of the fire spread. It spread extremely rapidly around the outside of the building, both vertically up the tower columns and laterally along the cladding above and below the window lines.[3] The fire moved across the eastern side of the building to the north face of the tower and into the majority of the other apartments in the tower block. The principal catalyst for the fire spreading horizontally and downward was dripping polyethylene emanating from the architectural crown of the building and from the spandrel and column panels. The polyisocyanurate (PIR) and phenolic foam insulation boards, located behind the ACM panels, contributed to the speed of the vertical flame spread.

Tragically, Grenfell Tower had a "stay put" fire policy in place which means that the building had been designed to contain any outbreak of fire within the individual flat where the fire ignited until the fire service arrived to extinguish the fire. However, in reality, there proved to be little resistance to the spread of fire making it extremely difficult to extinguish, which resulted in the death of 72 occupants.[4] The death toll was exacerbated by the volumes of smoke produced from the burning facade. The smoke appears to have entered the building through unprotected openings even before the contents of each affected flat were ignited and the high toxicity of the smoke from the facade is an important factor in the scale of the tragedy. Smoke inhalation is recognised as the most common form of death and injury from fire in the UK.[5] On exposure to the smoke, the victim becomes unconscious, and unless they are rescued from the building, the effect of the smoke is lethal.[6]

Public inquiry

The public inquiry was called by the Prime Minister, Theresa May, on 15 June 2017. It was chaired by a retired judge, Sir Martin Moore-Bick, and its primary purpose was to establish the facts of how the fire had occurred to prevent a similar tragedy happening in the future. Phase 1 of the inquiry considered what happened on the night of the fire. It commenced on 14 September 2017 and concluded on 12 December 2018.

A number of expert witnesses gave evidence at the Public Inquiry between 20 and 29 November 2018:

- Professor Luke Bisby spoke about the ignition of the facade materials and the external spread of fire.
- Professor Jose Torreo addressed the fire spread throughout the building.
- Dr Barbara Lane considered the fire protection measures within the building.
- Professor Niamh Nic Daeid gave evidence around the cause and spread of the fire in the flat of origin, Flat 16, and the spread of fire within and from Flat 16.
- Professor David Purser provided evidence on the production of toxic gases and the consequences of inhaling toxic gases in such circumstances, which were the likely causes of incapacitation and death at Grenfell Tower.

The Phase 1 inquiry report was published on 30 October 2019, and Moore-Bick concluded that

> there was compelling evidence that the external walls of the building failed to comply with Requirement B4(1) of Schedule 1 to the Building Regulations 2010, in that they did not adequately resist the spread of fire having regard to the height, use and position of the building. On the contrary, they actively promoted it.[7]

The Phase 2 inquiry commenced in February 2020 with the overarching aim of examining the events that lead to the fire. These events included the refurbishment of the tower, testing of the external cladding system, compliance with the building regulations and the role of central and local governments. The inquiry was suspended in March 2020 due to COVID-19 lockdown restrictions and did not recommence until 7 September 2020.

In the following sections, the evidence provided in Phase 1 of the inquiry is used to explain in more detail how the design of the refurbishment works, and the choice of materials used in those works contributed to the spread of fire and smoke throughout Grenfell Tower. The final section also reviews the "stay put policy" that was in place at the time of the fire and how this contributed to the tragedy.

Aluminium composite cladding material

The renovation of Grenfell Tower involved the building being externally clad with a ventilated rainscreen system designed to protect the building from direct rainfall with a cavity behind the outer skin to ensure that any rainwater was collected and drained away to prevent it from penetrating into the building. The rainscreen panels were known as "Reynobond 55 PE" Aluminium Composite Panels (ACP). Each panel consisted of two 0.5 mm thick aluminium sheets bonded to a 3 mm central plastic (polyethylene) core. Behind the panels was a layer of insulation fixed directly to the building. On the spandrels this, predominantly, consisted of two 80 mm layers of Celotex RS5000 PIR polymer foam with one 100 mm layer of the same insulation being affixed to the columns.

Before being installed, the panels were fitted into "cassettes" which were hung onto aluminium or steel supports affixed to the concrete structure. This left a 50 mm cavity between the inside face of the rainscreen panel and the outer face of the insulation to allow the ventilation and drainage of any rainwater that penetrated the gaps between the external cladding panels. Smaller cavities, with no design function, were also formed between the face of the insulation board and the columns. The problem with PIR foams is that they have a comparatively low time to ignition, and so they support rapid flame spread. In addition to this, the PIR foam in the cavities accelerated the spread of flame on adjacent materials.[8] See Figure 2, which is a horizontal section detail of the external wall system illustrating the position of the cavities between the insulation and the cladding.

At Grenfell, the vertical flame spread could also be explained by the cavities producing chimney effects, that is, the upward movement of hot air in enclosed vertical spaces. However, Professor Torreo concluded that given the relatively slow rate of vertical flame spread in comparison to other similar fires, the detailing of the external cladding system had only a minor effect. The important factor in the rate and extent of the flame spread was the composition of the materials used.

Cavity barriers were installed as part of the over-cladding system. They comprised an intumescent strip which expands in the event of a fire to seal the gap between the barrier and the rear of the cladding panels. However, cavity barriers were not provided to all the columns and no cavity barriers were present at the top of the columns nor at the head of the rainscreen cladding. Evidence presented to the inquiry outlined two major problems:

a   The cavity barriers were prevented from being continuous due to the presence of the cladding rails supporting the ACM panels.

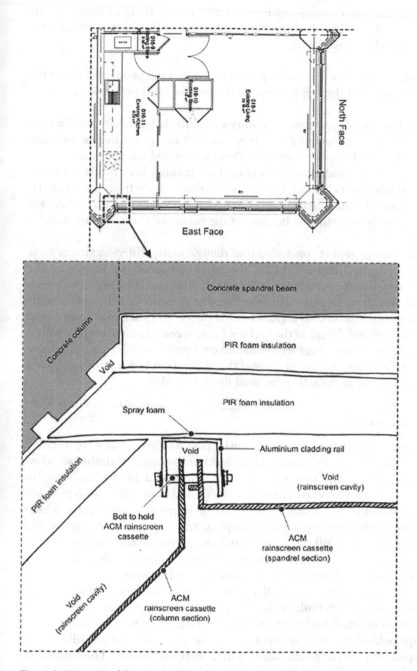

Figure 2 Horizontal Section Through the External Wall System.
(Credit Professor Bisby supplemental report [LBYS0000001] p. 43 Fig. 16.)

b   There were many instances of the cavity barriers being poorly fitted, so instead of the barriers being tightly abutted, there were gaps between them.

Evidence presented to the inquiry advised that the primary cause of the rapid spread of the fire was due to the presence of polyethylene.

> The ACM (aluminium composite material) product on Grenfell Tower incorporates a highly combustible polyethylene polymer filler which melts, drips and flows at elevated temperatures.
> … the fundamental assertion I am making here is that if a fire is ignited in a cladding system such as this is made from these materials under any circumstances, we have to expect it to spread quickly and catastrophically because of the nature of the materials used.[9]

The inquiry report concluded that though many different factors were involved, the main reason for the rapid spread of the flames up the building was the melting of ACM panels with polyethylene cores which provided the source of fuel for the fire. The presence of PIR and phenolic foam insulation boards located behind the ACM panels also contributed to the rate and extent of the vertical flame spread. Interestingly, whilst it was accepted that there were fundamental problems with the horizontal cavity barriers, the inquiry found that these defects were not significant with respect to the rate of vertical flame spread.[10]

Window units

As part of the major renovation works, new window units were installed on every floor of the tower whilst leaving the original timber window units in situ. The new window sills, jambs and head were made of unplasticised polyvinyl chloride (uPVC) and were glued into position as opposed to using mechanical fixings. Extruded polystyrene, which melts rapidly forming burning droplets, was covered in aluminium foil and was used to fill the gaps around the extractor fans and within the infill panels around the windows. Figure 3 shows a close-up of the completed window installation works to the higher floors of the tower.

The new units were smaller in size than the original windows, and instead of sitting flush with the original concrete, they were moved forward to finish flush with the new over-cladding system. This created gaps behind the new window units, some of which were filled with expanding polyurethane foam which is inflammable. The reduction in the size of the windows also created gaps between the sides of the

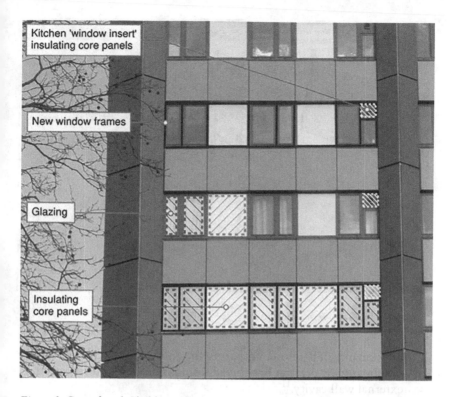

Figure 3 Completed Cladding Works.
(Credit: Dr Lane supplemental report [BLAS0000004] p. 35 Fig. 4.22.)

windows and the columns which was covered using a black EPDM (Ethylene Propylene Diene Monomer) membrane. EPDM is a combustible material and burns quite rapidly. The building regulations required cavity barriers to be located around the window to stop the spread of fire, but none were shown on the design drawings. Consequently, none were provided between the interior of the building and the cavity within the over-cladding facade, which effectively undermined the compartmentation of the building. Figure 4 illustrates the new window location in relation to the original window frame and the positioning of the EPDM weatherproof membrane.

In her evidence given to the inquiry, Dr Lane stated that

> In my opinion, once any localised fire occurred near a flat window – regardless of how that fire started – the majority of the construction materials around the window had no potential fire resisting

Figure 4 The New Window Unit.
(Credit: Dr Lane supplemental report [BLAS0000008] p. 16 Fig. 8.14 and p. 24 Fig. 8.25.)

performance. Therefore, no part of the construction had the ability to substantially prevent fire spread from inside the building into the external wall cavity.[11]

Once the uPVC window had deformed as a result of the temperature in the kitchen, it would have exposed a number of combustible materials including the 25 mm PIR insulation board and the EPDM membrane. The membrane would have burned through quite rapidly, and once that had happened, the flame was able to enter the back of the cladding cavity around the concrete column. The route of the potential fire spread is shown in Figure 5.

The inquiry reached the conclusion that it was the design of the refurbishment, together with the choice of materials and the method of construction, that allowed the kitchen fire to travel into the cladding with such tragic consequences, and it was highly likely that if a fire has started anywhere near a window, then it would escape from the flat and into the cladding.[12]

The crown

The refurbishment of the building also involved changes to the pre-cast concrete architectural "crown" at the top of the building. In the original

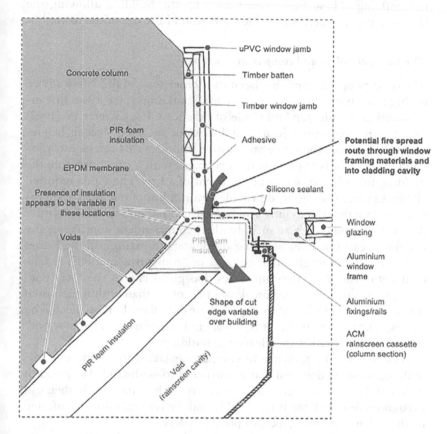

Figure 5 Potential Route of Fire Spread into the Cladding Cavity.
(Credit: Dr Lane report [BLAS0000008] p. 59 Fig. 8.65)

construction, this consisted of tapered pilasters at the tops of the columns and a ring of perforated freestanding concrete beam. As part of the refurbishment, the concrete columns and beams at the top of the tower were enveloped in a band of Reynobond 55 PE ACM cassettes, which had the visual appearance of C-shaped fins. These extended around the building above level 23. The fins and architectural crown were designed purely for aesthetic purposes, but crucially, exposed edges of PE were found everywhere in the crown.

The inquiry Chairman was satisfied that the principal means for the horizontal and downward spread of flame was the melting and dripping of burning polyethylene, emanating partly from the crown – and also from the spandrel and column panels – which ignited fires lower down the building. These fires moved back up the building allowing the flames to progress across each face and around the top of the tower.[13]

The stay put policy and compartmentation

The concept of "stay put" has been the cornerstone of fire safety advice for high-rise multi-occupancy residential buildings since it was first introduced in British Standard Code of Practice CP3: Chapter IV (1962) Part 1: Fire Precautions in flats and maisonettes over 80 ft in height. The Code sets out the active and passive measures that can be incorporated into HRRBs to provide the occupants with a safe route from the building to a place of safety in the event of a fire. The Code required all blocks taller than 80 ft to provide a minimum of one hour's fire resistance with the overarching aim being that each individual flat within the building would act as an individual compartment containing any fire for at least one hour enabling firefighters to extinguish the flames within the flat of origin. This concept of "compartmentation", in tandem with other fire safety measures, has given rise to the "stay put" strategy in which, at the outbreak of a fire, other than within their own flat, occupants are advised to stay in their own flats. Unless affected by flames, heat and/or smoke, it is considered safer for them to "stay put" in their own flat rather than leave the building.[14]

However, for the principle of compartmentation to be effective, the building must be designed and constructed to withstand the spread of flames and to provide egress and access to the building. Only then can firefighters deal with the blaze quickly and the residents directly affected by the fire are able to escape to a place of safety.

The National Fire Chiefs Council still supports the principle of a "stay put" strategy as they believe it has been proved safe and effective, over many years, for residents of purpose-built blocks of flats that are built and maintained correctly.[15] However, this position was challenged after the fire at Lakanal House, Southwark on 3 July 2009 where six people died after compartmentation within the flat of origin failed and the fire spread both externally and internally. The deceased had all been told to "stay put". The coroner responsible for the Lakanal House Fire inquest, Her Honour Frances Kirkham, wrote a letter to the Mayor of Southwark and to the Lord Chancellor in March 2013, making a number of recommendations related to fire safety in HRRBs. These recommendations included a request that residents in HRRBs be given clear guidance as to whether they should vacate their flat or stay put in the event of a fire breaking out in the building.[16] Despite this, and the concerns expressed by the coroner at the Lakanal House inquest, the "stay put" policy remained the default strategy for blocks of flats.

In 2011, less than two years after the fatal fire at Lakanal House, the Local Government Association, with inputs from the Department of Local Communities and Local Government and the Chief Fire Officers' Association, produced a report entitled *Fire Safety in Purpose Built Blocks of Flats*, which stated that:

> Some enforcing authorities and fire risk assessors have been adopting a precautionary approach whereby, unless it can be proven that the standard of construction is adequate for "stay put", the assumption should be that it is not. As a consequence, simultaneous evacuation has sometimes been adopted, and fire alarm systems fitted retrospectively, in blocks of flats designed to support a "stay put" strategy.
>
> This is considered unduly pessimistic. Indeed, such an approach is not justified by experience or statistical evidence from fires in blocks of flats.... Accordingly, proposals of fire risk assessors, and requirements of enforcing authorities, based on a precautionary approach (eg abandonment of a "stay put" policy simply because of difficulties in verifying compartmentation) should be questioned.[17]

The issue was further addressed in 2014. In the generic risk assessments for fighting fires in high-rise buildings issued by the Department for Communities and Local Governments, it advised that a suitable emergency evacuation plan should be devised by the responsible person under the provisions of the Regulatory Reform (Fire Safety) Order 2005, but that this plan could be overridden by the Incident Commander from the Fire Brigade, who should follow the evacuation plan unless the fire situation dictated otherwise. Where a "stay put policy" was in place for an affected building, it recommended that the occupants should self-evacuate when the fire, heat or smoke was adversely affecting them in their own property. This could have been as a result of a fire that started in their own property or from elsewhere in the building.[18]

Despite this guidance, derived from the tragic lessons learned from the Lakanal House fire, many Grenfell residents were advised by the emergency services to remain in their flats and became trapped by the smoke which spread upward within the single stairwell to the building. The London Fire Brigade (LFB) was heavily criticised by the Grenfell Inquiry chair, Sir Martin Moore-Bick, in the Inquiry's Phase 1 report which found that the "stay put" policy had become an "article of faith" within LFB and any decision to depart from this strategy was "unthink-able" to senior officers.

The inquiry concluded that the rapid failure of the compartmentation and the rate at which the smoke escaped into the lobbies and stairwell was of significant concern and showed that several active and passive fire protection measures had failed to operate correctly. The inquiry found that it was reasonable to assume that a number of factors including the failure of the protected fire escape route, the fire doors and the use of the lifts had caused the building to suffer "a total failure of compartmentation".[19]

Protected escape route and fire doors

The central enclosed stairwell was set within the thick concrete core of the building and was designed to be a protected escape route in the event of a fire. The lobbies to each floor should also have provided protection from the smoke and fire allowing access to the enclosed stairwell, but they failed to do so on the night of the fire for a number of reasons:

- The majority of the entrance doors to the flats did not close automatically as they should have done and residents gave evidence at the inquiry that many of the doors were faulty with missing closing mechanisms.
- The initial density of smoke in the lobbies is believed to have been caused by front doors being left open as the residents escaped from their flats.
- Tests conducted on the entrance doors after the fire showed that the doors resisted fire for 15 minutes as opposed to the 30 minutes recommended in Approved Document B.
- The smoke ventilation system, activated by smoke detectors, should have extracted smoke from the lobbies through the north and south smoke shafts using mechanical fans. However, the system was designed to offer smoke control to one floor only and could not clear smoke simultaneously from multiple floors.
- The protected shafts that ran from the smoke ventilation system may have exacerbated the spread of smoke due to the opening located between the lobby and shafts at every level.
- If the smoke control system had operated correctly, and the fire service had been able to take control, then it may have been possible to vent smoke from the lobbies on each floor of Grenfell Tower.[20]

As many of the doors had been destroyed in the fire, there was uncertainty about their condition prior to the start of the fire. However, the inquiry thought that it was safe to conclude that some of the flat entrance doors had failed to control the spread of smoke and fire effectively and that these deficiencies allowed smoke to spread at an early stage in the fire.[21]

Lifts

Both lifts had been refurbished between 2005 and 2006. The works comprised replacement of each of the lift cars and renovation of the lift motor room. However, neither of the lifts had been upgraded to become a firefighting lift nor, on the night of the tragedy, were the firefighters able to operate the override mechanisms which should have allowed them to take control of the lifts and use them to assist in their firefighting duties. Unfortunately, this meant that some residents were able to use the lifts which quickly filled with smoke leading, in some cases, to fatal consequences.

Recommendations of the Phase 1 inquiry[22]

Rainscreen cladding

It was evident that the use of the ACM rainscreen cladding and the combustible insulation was the main reason for the quick spread of the fire. Since the fire, the UK Government has officially recognised that there are 478 residential buildings over 18 m in height in England with ACM cladding in place.[23] Of the 328 buildings for which data is available, 316 do not comply with the building regulations and half of these buildings are social housing.[24] Quite rightly, the residents living in these buildings are concerned for their safety. The Government has directed that ACM external wall systems must be removed from HRRBs, and they have provided support for local authorities via the Building Safety Fund to undertake emergency works to replace ACM cladding.[25] The Inquiry also highlighted that in the wake of the part played by the architectural crown in the spread of fire at Grenfell, particular attention should be directed towards the use of combustible materials in decorative features.

Fire doors

Ineffective fire doors with missing or broken self-closing devices allowed smoke and toxic gases to spread through the building more quickly than they should have. Moore-Bick recommended that owners and managers of every residential building containing separate dwellings, whether HRRBs or not, should carry out inspections of all fire doors every 3 months to ensure they comply with the applicable legislative standards and are fitted with effective self-closing devices in good working order. The recommendation was extended to ensure that the front entrance doors to individual flats in HRRBs should comply with current standards.

Sprinklers

The inquiry report did not make any recommendations about the use of sprinklers, as it was not proven that a sprinkler system would have prevented the fire from spreading at Grenfell Tower. It was, however, noted that such systems have an integral part to play with respect to fire safety measures and the report reiterated the advice provided by the coroner, who presided over the Lakanal House fire inquest, that the government should encourage housing providers responsible for HRRBs containing multiple domestic premises to consider fitting sprinkler systems.

Lifts

On the night of the Grenfell tragedy, firefighters were unable to operate the lift override mechanism. The inquiry recommended that the owners/managers of HRRBs should be statutorily obliged to undertake the regular inspections of all lifts and this should include testing of the override mechanism designed to allow firefighters to take control of the lifts in emergencies. The inquiry further recommended that the results of these inspections and tests should be reported to local fire and rescue services on a regular basis.

As-built drawings

Moore-Bick acknowledged that the Fire Brigade were not unduly hampered by the absence of accurate floor plans in its attempt to fight the fire at Grenfell Tower. However, the view was taken that the lack of floor plans could have far more serious consequences, and therefore, the inquiry recommended that the owner of every HRRB be required, by law, to provide the local fire and rescue services with up-to-date plans, in both analogue and digital formats, of all floors of the building identifying the location of key fire safety systems. Each building should also contain a premises information box, the contents of which must include copies of the up-to-date plans and information about the nature of any lift that could be used by the fire and rescue services.

Phase 2 inquiry

At the time of writing this book, the Phase 2 inquiry had been delayed by the effects of the coronavirus pandemic and the inquiry hearings were ongoing. However, the evidence given to date has revealed that Arconic, the firm that manufactured and sold the external combustible ACM cladding cassettes used on Grenfell, were aware that the cladding was combustible. The Reynobond panels came in two types – cassette and rivet-fixed – and in 2004 a fire test showed that there were vast differences between the two forms with the cassette version performing significantly worse than the rivet-fixed. The test data was not shared with the relevant certification bodies, and both types were sold under the same fire safety accreditation, even though it had been shown that the rivet-fixed had significantly better fire resistance performance. A British Board of Agrément (BBA) certificate, issued in 2008, made no distinction between the two versions of Reynobond PE, and even though the cassette variant did not achieve any rating in the 2004 test, it was presented as having a Euroclass B rating. This rating was used as the basis for the BBA giving the cassette variant a UK class 0 rating which meant that it was safe for use on buildings over 18 m in height.[26]

Arconic also knew that there had been several fires in the Middle East in 2013, where the buildings had been clad in ACM of the same type as Arconic's Reynobond PE panels.[27] Arconic had been reliably informed that the ACM cladding could behave like a "chimney which transports the fire from bottom to top or vice versa within the shortest time". They refused to withdraw the Reynobond PE panels from the marketplace, even though a competitor, 3A, had withdrawn a similar Alucobond product. At the time when Arconic was negotiating the Grenfell contract, they sold two types of Reynobond cladding. The first was fire retardant, known as Reynobond FR, and had a 70% mineral core. The other was the Reynobond PE, which had a polyethylene core and was cheaper, but combustible. Arconic's profit margin was greater for the more commercially popular Reynobond PE cladding. The evidence revealed that no consideration was given to the fire safety consequences of continuing to sell Reynobond PE.[28]

In 2014, the fire rating of Reynobond PE was downgraded from Euroclass B standard, which recommended that it was safe to use on high-rise buildings, to the much lower-class Euroclass E which meant that it should not be installed on high-rise buildings. The downgrading was effected after the polyethylene core had burnt ferociously in a fire test.

After the downgrade, Arconic stopped selling Reynobond PE in France but continued to sell the product for architectural use in the UK and only halted supply to the UK 12 days after the Grenfell Tower fire. Arconic's view put forward at the Inquiry was that the firm did not believe that Reynobond PE was inherently dangerous, but that it had certain characteristics that, if it was used "incorrectly", would increase the risk of fire spread because the polyethylene core would burn if it was exposed to a heat source. Arconic's position regarding the fire at Grenfell was that the product had been used incorrectly in conjunction with other materials and products and that the insulation made by others caught alight first.[29]

Similarly, the two insulation suppliers, Celotex and Kingspan, have been accused of ignoring the results of testing regimes undertaken to check fire safety, and of circumventing regulations by using misleading marketing strategies relating to the performance of their products. Celotex FR5000 PIR insulation was originally specified for use in the Grenfell Tower refurbishment in 2012, but it failed a Building Research Establishment fire test in April 2014. The product was rebranded as RS5000 and passed the BS8414 fire test. But the test was manipulated as the insulation system presented for the fire test included non-combustible magnesium oxide board which improved the fire performance rating ensuring that the test was passed.

The "new" RS5000 was specified and used as the main insulation in the refurbishment works at Grenfell. RS5000 had been granted a Local Authority Building Control Certificate (LABCC), declaring it to be safe to be used on buildings above 18 m in height, as it had achieved a Class 0 rating. The certificate may, however, have been issued in error, due to a misconception that a Class 0 rating meant that a product was of limited combustibility when, in fact, the rating only refers to the fact that a product adequately resisted the surface spread of fire. Similarly, Kingspan, who made the rest of the insulation, carried out tests that allegedly involved concealing components to assist in achieving the desired rating and/or using materials that were not as explicitly described in the test reports. Both Celotex and Kingspan dispute the allegations that have been made against them. Celotex has stated that Arconic misled the industry regarding the safety of the Reynobond cladding panels and that the construction professionals involved in the refurbishment works at Grenfell Tower failed to follow the building regulations. Kingspan has cited Arconic's panels as the major issue and argues that even if non-combustible insulation had been used, the outcome of the fire would have been no different. These will be crucial issues that the Inquiry will determine upon.[30]

Summary

The narrative of Chapter 1 exemplifies the nature of the overall systemic failure that led to the tragic fire at Grenfell Tower. The failure to learn from the lethal and no less tragic fire at Lakanal House in 2009, and the manipulation of fire safety test results by materials manufacturers are prime examples of why Dame Judith Hackitt put forward such a pessimistic view of the UK construction industry. She condemned the industry for displaying characteristics of ignorance and indifference with a primary motivation to do things as quickly and cheaply as possible as opposed to delivering quality homes. Hackitt concluded that safety is not always prioritised and the vagueness and lack of understanding around regulations allowed some actors in the industry to game the system.

Perhaps the most crucial lesson to come from the Grenfell tragedy was the role of residents and the need to listen to them. Not only are they entitled to have a voice in how their homes are managed, but they are often more aware of what happens in their buildings than the property management team. Rather than viewing tenants as problems to be managed, they should be empowered by legislation to become part of the risk-mitigation strategies to establish safety priorities for their buildings and their homes.

Notes

1 The description of Grenfell Tower and the refurbishment works has been taken from the Grenfell Tower Inquiry: Phase 1 report Vol 1 Chapter 3. Moore-Bick, M (2019) Grenfell Tower Inquiry: Phase1 Report Overview. Open Government Licence.
2 Department for Communities and Local Governments (2006) A Decent Home: Definition and Guidance for Implementation. HMSO, Holborn, London, UK.
3 Lane, B (2018) Grenfell Tower Fire Safety Investigation. Phase 1 Report – Section 9. Routes for fire spread out through the window openings pp 9–46 Dr Barbara Lane report – section 9 (Phase 1 – supplemental).pdf (grenfelltowerinquiry.org.uk) (Accessed 31/12/2020).
4 Gorse, C & Sturges, J (2017) Not What Anyone Wanted: Observations on Regulations, Standards, Quality and Experience in the Wake of Grenfell. *Construction Research and Innovation* Vol 8(3) pp 72–75. https://doi.org/10. 1080/20450249.2017.1368260
5 UK Fire Statistics (2017) and Preceding Editions, Home Office, London, https://www.gov.uk/government/collections/fire-statistics (Accessed 01/06/2020).
6 Mckenna, ST, Jones, N, Peck, G, Dickens, K, Pawelec, W, Oradei, S, Harris, S, Stec, AA, & Hull, R (2019) Fire Behaviour of Modern Façade

Materials – Understanding the Grenfell Tower Fire. *Journal of Hazardous Materials* 368 pp 115–123.

7 See page 13 of Moore-Bick, M (2019) Grenfell Tower Inquiry: Phase 1 Report Overview. Vol 1. Open Government Licence.

8 See pages 33–37 of Moore-Bick, M (2019) Grenfell Tower Inquiry: Phase 1 Report Overview. Vol 1. Open Government Licence.

9 Bisby, L (2018) Expert Evidence at the Grenfell Inquiry November 2018.

10 See pages 558–559 of Moore-Bick, M (2019) Grenfell Tower Inquiry: Phase 1 Report Overview. Vol 4. Open Government Licence.

11 Lane, B (2018) Grenfell Tower Fire Safety Investigation. Phase 1 Report – Section 9. Routes for fire spread out through the window openings pp 9–46.

12 See page 538 of Moore-Bick, M (2019) Grenfell Tower Inquiry: Phase 1 Report Overview. Vol 4. Open Government Licence.

13 See page 558 of Moore-Bick, M (2019) Grenfell Tower Inquiry: Phase 1 Report Overview. Vol 4. Open Government Licence.

14 See page 27 of Moore-Bick, M (2019) Grenfell Tower Inquiry: Phase 1 Report Overview. Vol 1. Open Government Licence.

15 National Fire Chiefs Council (2020) Simultaneous Evacuation. https://www.nationalfirechiefs.org.uk/Simultaneous-evacuation-guidance (Accessed 2/6/2020)

16 Barling, K (2017) We've Been Here Before. *British Journalism Review* Vol 28(3) pp 30–35. doi:10.1177/0956474817730766

17 Local Government Association (2011) *Fire Safety in Purpose Built Blocks of Flats* LGA p 27, UK.

18 Department for Communities and Local Governments (2014) *Generic Risk Assessments 3.2 – Fighting Fires in High Rise Buildings* (p 28). HMSO, Holborn, London UK.

19 See page 571 of Moore-Bick, M (2019) Grenfell Tower Inquiry: Phase 1 Report Overview. Vol 4. Open Government Licence.

20 The Chartered Institution of Building Services Engineers (2018) What Went Wrong with the Smoke Ventilation at Grenfell Tower. *CIBSE Journal* June 2018. What Went Wrong with Smoke Ventilation at Grenfell Tower – *CIBSE Journal.*

21 See page 570 of Moore-Bick, M (2019) Grenfell Tower Inquiry: Phase 1 Report Overview. Vol 4. Open Government Licence.

22 See pages 771–774 of Moore-Bick, M (2019) Grenfell Tower Inquiry: Phase 1 Report Overview. Vol 4. Open Government Licence.

23 MHCLG Building Safety Programme: Monthly Data Release – August 2018, and preceding editions.

24 Mckenna, ST, Jones, N, Peck, G, Dickens, K, Pawelec, W, Oradei, S, Harris, S, Stec, AA, & Hull, R (2019) Fire Behaviour of Modern Façade Materials – Understanding the Grenfell Tower Fire. *Journal of Hazardous Materials* 368 pp 115–123.

25 Phillips, S (2021) *The Valuation of High-Risk Residential Buildings and the Role of EWS1 The Journal of Building Survey, Appraisal & Valuation.* Vol (4) pp. 305–314

26 Grenfell Tower Inquiry (2021) Transcript: Day 89 11 February 2021 OPUS 2-Official Court Reporters Transcript 11 February 2021.pdf (grenfelltowerinquiry.org.uk) (Accessed 17/2/2021).

27 British Broadcasting Corporation (2013) 'Towering Inferno' Fears for Gulf's High-Rise Blocks. 'Towering inferno' fears for Gulf's high-rise blocks – *BBC News* (Accessed 16/2/2021).

28 Lowe, T (2021) Grenfell Inquiry Round-up: What We Learnt from Kingspan and Arconic. Building Magazine Grenfell Inquiry Round-up: What We Learnt from Kingspan and Arconic | News | Building (Accessed 16/2/2021).

29 Lowe, T (2021b) Arconic Stopped Offering French Customers Grenfell Cladding Because of Fire Safety Concerns. Building Magazine Arconic Stopped Offering French Customers Grenfell Cladding Because of Fire Safety Concerns, Inquiry Hears | News | Building (Accessed 16/2/2021).

30 Dunton & Lowe, T (2020) Grenfell Inquiry: What We Learned from ex-Celotex Staff This Week. 20 November 2020. Building Grenfell Inquiry: What We Learned from ex-Celotex Staff This Week | News | Building (Accessed 17/2/2021).

2 The Hackitt Review

Background

Following the tragic fire at Grenfell Tower, fire assessment audits were undertaken of tower blocks throughout the UK. The audits revealed a number of structural safety issues with the design of cladding systems.

- In Glasgow, materials had fallen from tower blocks as a result of the way their cladding system was designed and installed.[1]
- Residents from five tower blocks on the Chalcots Estate, Camden, were temporarily evacuated after the discovery of a number of fire safety issues.[2]
- Structural safety faults were uncovered in four buildings on the Ledbury Estate, Southwark.[3]

In the light of these revelations and others like them, Dame Judith Hackitt was asked by the Home Secretary and the Secretary of State for the Department of Communities to conduct an independent review of building regulations and fire safety with a particular focus on their application to HRRBs. The review was separate from the Grenfell Tower Inquiry and it did not seek to identify the cause of the fire. Instead, the review focused on establishing a robust, regulatory regime for the future which would provide assurance to residents that the buildings they live in are safe and will remain safe.[4]

Dame Judith is a chemical engineer who worked in the chemicals manufacturing industry for 23 years before joining the Chemical Industries Association (CIA) where she was the Director General from 2002 to 2005. Dame Judith was a health and safety commissioner from 2002 to 2005 and was the Chair of the Health & Safety Executive from 2007 to 2016. She is a Fellow of both the Royal Academy of Engineering and the Institution of Chemical Engineers.[5]

DOI: 10.1201/9781003092803-2

Dame Judith's considerable experience in the chemical engineering industry has strongly influenced the narrative of the Independent Review as it highlights that the basic disciplines of proper control and review of change management that exist in chemical engineering are sadly lacking in the built environment sector. The Review document reveals that buildings are often completed without a comprehensive record of what has been built, and consequently key safety features of the building are lost. This lack of accurate record-keeping means that it is very difficult for the managers of multi-occupancy HRRBs to do their jobs effectively and to maintain the safety of their building particularly when there is no thorough safety review system in place for future changes to the building.[6]

The interim report

The review team began work in August 2017 and sought wide-ranging views and evidence from across the industry. This included those working in the built environment sector, regulators and residents of HRRBs. It was clear that many of these people knew the system was not working effectively. Hackitt published an interim report in December 2017, which produced a full map of the regulatory system highlighting a significant level of complexity that allowed people to take short cuts or game the system. The report concluded that the current regulatory system for ensuring fire safety in high-rises and complex buildings is not fit for purpose and that a radical overhaul was required of both the culture of the construction industry and the effectiveness of the regulators.[7]

The reasons for this failure of the regulatory system were multiple. The guidance provided was complex and impenetrable for the user, and roles and responsibilities were unclear. There was no differentiation in the competency requirements for those who worked on relatively simple, straightforward buildings and for those who worked on high-rise complex buildings. Compliance was weak and enforcement and sanctions were virtually non-existent, while the system of product testing did not provide meaningful quality assurance.[8] The interim report identified six main areas that would be addressed in the final report with some initial direction of travel[9]:

1   **Regulation and guidance** – These should be simplified and unambiguous. The rules for ensuring that high-rise and other complex buildings are built safe and remain safe should be more risk-based and proportionate. Those responsible for high-rise

and complex buildings should be held to account to a higher degree.

2   **Roles and responsibilities** – Primary responsibility for ensuring that buildings are fit for purpose must rest with those who commission, design and build the project. Responsibility and accountability must rest with clearly identifiable senior individuals over the life of the building

3   **Competence** – There is a need to raise levels of competence and establish formal accreditation for those engaged in fire prevention.

4   **Process, compliance and enforcement** – There needs to be a "golden thread of information" for all high-rise residential and complex buildings. The original design intent and any subsequent changes or refurbishment must be recorded and properly reviewed with regular reviews of overall building integrity. It is envisaged that the "golden thread" will be held as a digital record using BIM as the industry wide digital platform.

5   **Residents voice** – Residents need reassurance that there is an effective system in place to maintain safety in their homes. There must be a clear, quick and effective route for residents' concerns to be heard and addressed.

6   **Quality assurance and products** – All products must be correctly tested, certified and installed in accordance with the relevant quality control standards. The marketing of these products being clear and easy to interpret.

Six working groups were formed to develop proposals in each of the six areas identified in the interim report. The groups were also tasked with looking for examples of good international practice which could be learned from. It became clear that this was a wider problem within the UK than first appreciated and the Scottish Government commissioned its own review of its building standards, compliance and fire safety.[10]

The final report

The final report was published on 17 May 2018 with an overarching objective to address the system failure identified in the interim report through a new regulatory framework for the construction and management of high-rise residential multi-occupancy buildings, which would drive a real culture change and different behaviours throughout the system.[11]

Systemic failure

Hackitt was scathing in her view of the construction industry and described it as *"an industry that has not reflected and learned for itself, nor looked to other sectors"*[12] and she set out four key issues that underpinned the systemic failure:

1 **Ignorance** – The regulations and guidance notes are not always read by key actors within the supply chain, and when they are, the guidance may be misunderstood and misinterpreted.
2 **Indifference** – The primary motivation is to do things as quickly and cheaply as possible rather than to deliver quality homes which are safe for people to live in. The residents and their concerns are often ignored. Safety is not always prioritised, and the vagueness of regulations and guidance is used to game the system.
3 **Lack of clarity on roles and responsibilities** – There is ambiguity over where responsibility lies, and this precludes robust ownership of accountability.
4 **Inadequate regulatory oversight and enforcement tools** – The size or complexity of a project does not seem to inform the way in which it is overseen by the regulator. Where enforcement is necessary, it is often not pursued. Where it is pursued, the penalties are minimal and are an ineffective deterrent.

In Hackitt's view, these four issues had combined to create an inherent cultural problem within the industry where ignorance and indifference have caused a "race to the bottom" and a lack of focus on delivering the best-quality building possible so that the residents are safe and feel safe.

New systems-based approach to building safety

The experience Dame Judith had gained as a chemical engineer with regard to process safety management and operational integrity strongly influenced the content of the final report and its proposals for a new regulatory framework. Considerable emphasis was placed on the need to change the whole system of how complex high-rise buildings are developed and managed to ensure that the built environment sector takes full responsibility for delivering and maintaining buildings which are safe for people to live and work in.[13]

The Hackitt Report made 53 recommendations concerning the formation and operation of a proposed new regulatory framework which will benefit building standards across the board. The report recognises that HRRBs are comprised of many inter-connected systems, some physical, such as fire compartmentation, and some system-based, such as the responsible persons who ensure the fire safety of buildings. Hackitt made it clear that although each individual recommendation was crucial, they could not, in isolation, achieve the systemic change required. Hackitt also recognised that implementing all the recommendations would take time. As whilst some could reasonably be introduced in the short term, some would require primary legislation to be enacted. With these two things in mind she urged that two things should happen. Firstly the industry should start "living" the cultural shift required and secondly the Government should develop "joined-up thinking" to implement a plan to deliver all the report recommendations. Hackitt also stated that the recommendations should apply not just to HRRBs but should be considered for all buildings where residents are at an enhanced risk of harm from fire. This should include other multi-occupancy residential buildings below ten storeys and buildings providing accommodation where people sleep, such as hospitals, care homes, prisons and halls of residence.

The report recommends the introduction of a safety case approach to high-risk complex buildings with the regime overseen by a competent regulator. Drawing on the recent evolution of health and safety law, particularly the assignment of specific, legally enforceable duties, on designated duty holders,[14] Dame Hackitt recommends an overarching accountability throughout the development and management processes with specific duties for those responsible for the procurement, design, construction and management of buildings with the relevant duty holder being clearly identified at each stage of the process.

The recommendations of the report can be summarised into nine main areas:

The new regulatory framework

The focus of the new regulatory framework set out in the Hackitt Report is with respect to both new and existing HRRBs of ten storeys or more because the likelihood of fire is greater than those with fewer storeys. There are an estimated 2,000–3,000 HRRBs in the UK, and new HRRBs should be identified by local planning authorities.[15]

It was recommended that the new regulatory framework be overseen by a single regulator, termed in the report as the new "Joint Competent Authority" [JCA] whose purpose would be to oversee safety within HRRBs throughout their complete life cycle. It was envisaged that the JCA would combine the expertise of the Local Authority Building Control services (renamed as "Local Authority Building Standards" in recognition of the enhanced role they would fulfil), Fire and Rescue Authorities and the Health and Safety Executive. It was proposed that this new approach would provide the framework for these organisations to work together and utilise "joined-up thinking" to assess building safety more rigorously.[16]

The report also recommends a move away from the more prescriptive sections of the Approved Documents to ensure that people think more holistically about the requirements of the Building Regulations. It is envisaged that a new overarching Approved Document will be developed, which treats each building as a system ensuring that designers and specifiers undertake a holistic analysis which must consider how their proposed work will affect the building's interdependent systems.

Duty holders and their responsibilities

The interim report warned that a lack of clarity around roles and responsibilities in ensuring building safety was one of the prime issues that paved the way to the Grenfell Tower tragedy. In response, the final report focused on an outcomes-based approach to building safety that attaches importance to identifying duty-holders and assigning key responsibilities and accountability to these roles rather than expecting that a building has been constructed safely by relying on prescriptive rules and complex guidance. It is envisaged that to avoid confusion, the key roles will reflect those identified in the Construction Design and Management (CDM) Regulations 2015. These are set out in Table 1.

It is important to note that whilst there is a degree of overlap between the roles the existing roles defined by the CDM Regulations, the Hackitt Report sets out a new regime of responsibility and accountability for duty-holders.[17] The report also states that the specific responsibilities for complying with building regulations will be allocated to three duty-holders: the client, the principal designer and the principal contractor.

CLIENTS

The client organisation will be required to initiate the digital record and fire emergency file (FEF) for all buildings of over ten storeys. These will be handed over to the principal designer and principal contractor to take forward and develop. In addition to being responsible for assigning competent people to prioritise building safety, it is also expected that client organisations broker contracts with contractors that specifically state that safety requirements must not be compromised for cost reduction.[18] The report details how clients are expected to manage

Table 1 Key Roles under the CDM Regulations (Credit: Hackitt Final Report, p. 33).

Key roles under CDM regulations	Is this role critical in ensuring a focus on building safety?	Why is this role critical?
Clients	✓	Develops and maintains a sense of ownership and responsibility for building safety and regulatory compliance. Identifying the client's responsibilities at the outset will ensure a greater degree of ongoing engagement.
Principal designers	✓	Maintains the ownership concept on behalf of the client to ensure that Gateway Points are observed, and key players are engaged appropriately.
Designers	✓	Ensures accountability and helps create an audit trail in respect of any design changes that can be followed back through the Principal Designer and ultimately to the client.
Principal contractors	✓	Assumes primary ownership throughout the construction phase, and especially at handover to the occupation and maintenance phase.
Contractors	✓	Ensures accountability and helps to create an audit trail to ensure that any on-site changes can be followed back through the Principal Contractor and ultimately to the client.

developments so that they deliver against core objectives on building safety and to run procurement processes with dedicated time and resources to ensure that this is achieved. Clients will also need to be able to confirm at the completion of the works that all works do comply with building safety requirements and the Hackitt Report envisages that this may mean that clients will choose to rely on a clerk of works to monitor these issues on site.

DESIGNERS

The principal designer will be expected to demonstrate how the key building safety requirements for HRRBs will be met, as well as providing the regulator with full plans detailing how the core risks to building safety identified have been assessed and how the risk has been managed. They will be responsible for making sure that the project information management systems have been correctly updated. The principal designer will be one of the signatories required at the works completion stage to confirm that the building is safe and fully complies with the building regulations. In conjunction with the principal contractor, the principal designer will be responsible for finalising the digital record and FEF and ensuring that the information held in both is complete and accurate.

CONTRACTORS

The new regime will require contractors to take on a greater responsibility for building safety throughout the construction process. Contractors will have to not only maintain a digital record of specifications but will be required to amend these specifications to record and reflect any changes that occur during the construction process. This is something that contractors have not, historically, done well. The principal contractor will be responsible for the planning and management of key safety objectives throughout the construction phase. This will include preparing a construction control plan which sets out how compliance with building regulations and building safety will be achieved and how any changes or variations to the contract will be controlled and recorded. Along with the client and the principal designer, the principal contractor will be required to confirm by signature that the completed works fully comply with building regulations. Contractors will also be tasked with handing over the digital record and the FEF to the future building owner.

The Hackitt Report also sets out how the JCA will oversee and interact with the duty holders by:

- Retaining a database of HRRBs with details of key duty holders for those buildings
- Undertaking interventions to ensure that duty holders' competence and compliance controls are in place
- Conducting safety case reviews of duty holders' actions to ensure that the necessary competence and compliance controls are in place
- Ensuring that duty holders make building improvements to reduce building safety risks
- Advising duty holders on safety concerns
- Requesting tests of construction products, including annual tests from the appropriate testing bodies
- Working on a full recovery basis which means that the JCA will charge fees with more complex schemes paying a proportionately higher fee

The three gateways

The regulatory oversight of the duty holders' activities should be strengthened by the creation of a clear set of "Gateway Points" at key stages in a building's life cycle. The gateway intervention points allow the JCA to ensure that the duty holders are complying with the regulations at each key stage. The first gateway point is to satisfy the JCA that a proposed building is accessible by the fire service before the local planning authority will provide a determination on any planning application. The second gateway point is that prior to the commencement of any building works, the duty holder must satisfy the JCA that in addition to the plans meeting all the building regulation requirements, the Full Plans demonstrate that the key building safety risks are understood and will be managed. At the third gateway, the relevant duty-holder must satisfy the JCA that the signed off design has been followed during the construction phase and that any changes have been verified as acceptable. All key documentation must be handed over, with the residents' engagement strategy in place before occupation of the building can be commence.

These gateway points should combine to focus construction professionals on producing high-quality designs with associated high-quality delivery. The creation of a system in which duty- holders must meet the necessary requirements at each of these three stages before they can obtain permission for land use, commence building work or occupy the finished building should drive what Hackitt terms "the right behaviours".[19]

Greater powers of enforcement

The overarching objective of the Hackitt recommendations was to create a proportionate and effective enforcement system with a wider and more flexible range of powers that will result in the creation of reliably safe buildings from the outset. There are to be serious penalties for actors who choose to "game the system" and place the residents at risk. The stronger enforcement powers should be aligned with the Health and Safety at Work Act and the JCA and the LABS should be given additional powers to issue improvement and prohibition notices. Time limits for bringing prosecutions against duty holders should be increased to five or six years for "major deficiencies" in building requirements which are identified at a later date.[20]

The report also acknowledged the problems that have been created by the privatisation of building-control and it was proposed that though approved inspectors can still be used any regulatory oversight they provide must be independent of developers and contractors. This is to prevent a perpetuation of the existing situation which has been described as *"builders choosing their own regulators"*.[21]

Higher competence levels

The report identifies poor consistency in the processes for assuring the knowledge skills and behaviours of actors working on HRRBs as a major flaw in the current regulatory system. This is largely due to a fragmented approach to teaching and learning, with different competence frameworks even within a single discipline. There may be a lack of professional qualifications for operatives and even where qualifications do exist, there is no coherent way to evidence them so that they are clearly understood and recognised by those operating in the system.[22]

The construction sector will have to demonstrate much more effective leadership to ensure building safety competence amongst the core roles. This leadership must include the formation of an overarching body to provide scrutiny of the competence requirements. The far-reaching aim is to move the industry towards developing a system wherein the ownership of technical guidance rests with the industry itself as the intelligent lead in delivering building safety. This system should be agile enough to ensure that the provision of technical guidance keeps pace with changing practices.

Improving product testing and information

The report states that products used throughout the life-cycle of a building have a critical bearing on the safety of the building. Clearly products used in all buildings should be tested but it is crucial that this is accurately undertaken and reported for products used in HRRBs. It was not for the report to redesign the entire testing system but a more transparent, effective specification and testing regime must be developed. It must specifically target products that are put together for use as part of a system. Clear statements should be provided on what systems can and cannot be used for. The scope of the testing regime and the resulting implications must be clearly communicated to actors within the industry using plain, consistent and relatively non-technical language to ensure that the right products are being used in the right combination situation and context. This will support those who are responsible for delivering safe buildings in the discharge of their duties. Additional testing stations should be established and certified and testing methods and standards should be subject to a periodic review process.[23] Hackitt further recommended that assessments in lieu of tests, such as desktop studies, should be used sparingly. When used they must be undertaken by people who are qualified and/or competent and the study should be meticulously documented.

Procurement

The interim report highlighted the need for a cultural shift in the procurement process and recommended a move away from contractor selection based purely on financial criteria and towards prioritising the safety of buildings and high standards of construction quality. This recommendation was developed further in the final Hackitt Report which recommends that for HRRBs clients and principal contractors should devise contracts that specifically state that safety requirements should not be compromised by saving on costs.

In addition to this, the tender bid process must prioritise building safety and take a best value approach, balancing upfront capital cost against quality and effectiveness. All the safety requirements for a proposed HRRB should be effectively tested out during the tendering process and as part of the bid review which allows all parties to a contract to provide constructive feedback on these requirements. Efficiency and increased productivity are to be encouraged but should not be achieved by using unsuitable or cheap materials.[24]

The golden thread of building information

The Hackitt Report recognised that there are, currently, significant issues around the production and handover of building information by those responsible for design and construction works to the Freeholder or duty-holder in the occupation phase. This lack of accurate and complete building information can pose problems and make it difficult to determine whether any changes which may impinge on the building safety strategy have been made between the original design and the completed works. Similarly, if the building owner does not hold up to date information about their building, then it will be difficult for safety to be effectively managed across its life-cycle.

The report identifies four "key information products" which are deemed to be essential to the oversight of building safety. These are the digital record of the golden thread of information, the fire and emergency file (FEF), full plans and the construction control plan. Hackitt's recommendation is that the creation, maintenance and handover of this essential information must be an integral part of the legal responsibilities of clients, principal designers and principal contractors undertaking works on HRRBs.

The residents voice

The Hackitt Report places significant emphasis on ensuring that residents both are safe and feel safe in their own homes. Achieving this involves giving residents a greater voice and ensuring that not only do they have access to information about their homes but there are clear channels of communication which encourage complaints or misgivings to be routinely made known and investigated. The report also recommends empowering residents through tenants' panels and residents associations to play a meaningful role in residential management decisions which affect their homes. Historic information such as Fire Risk Assessments and Health and Safety Audit reports should be readily available to residents if requested. The final recommendation is that a new independent body or ombudsman should be formed to allow residents to escalate concerns that have not been reasonably dealt with by a specific duty-holder.

Summary

The main criticism of the Hackitt Review was that it failed to recommend an immediate ban on the use of combustible cladding material on HRRBs. Instead, the review had recommended that the Government should adopt a more robust and auditable product-testing regime and that external cladding should be retested at least once every three years to ensure that it was safe to be left in situ. The RIBA described this as "a major missed opportunity" and went as far as to say that the review had failed to act on the urgent requirement to protect life safety as it "offered no changes whatsoever to the actual regulations or baseline guidance".[25] The review was also criticised for failing to address the issue of banning the use of desktop studies and for not calling for the installation of sprinkler systems in HRRBs.

Dame Hackitt's response to these criticisms was to reiterate that the core issue was not about defective cladding and that simply adding more prescriptive measures to the existing fundamentally flawed system would not address the root causes of the problem. The UK construction industry must understand that the overarching problem with building safety is systemic. The current system is too complex, lacks clarity on who is responsible for what and there is insufficient regulatory oversight, coupled with ineffective enforcement. To address these problems, the industry has to move away from siloed thinking and begin to view buildings as a system that requires layers of integrated safety protection. If a change is made to one component system, then the duty holders must assess the repercussions this alteration will have on the other interdependent systems and on the building as a whole. If the industry continues to focus only on the individual problems such as the type of external cladding used on an HRRB, then the basic intent of fire safety will have been lost. The Hackitt Review called upon the construction industry to start living these changes now and challenged the Government set out a clear plan for the implementation of the review's recommendations. The Government's response to the report is the subject of Chapter 4.

Notes

1 Barnes, S (2017) Cladding Can "Fail" in Strong Winds, Javid Reveals. Inside Housing. Inside Housing - News - Cladding can 'fail' in strong winds, Javid reveals (Accessed 3/1/2021).
2 Camden Newsroom (2017) Evacuation of Chalcots Estate – Leaders Statement. 01:24 Evacuation of Chalcots Estate - Leader statement (camden.gov.uk) (Accessed 3/1/2021).
3 Jessel, E (2019) Residents Fear Demolition of Peckham's Ledbury Estate Following Structural Report. *Architects Journal*. Residents fear demolition of Peckham's Ledbury Estate Following Structural Report (architectsjournal.co.uk) (Accessed 3/1/2021).
4 Ministry of Housing, Communities and Local Government (2018) Radical Reform of Building Regulatory System Needed. 'Radical Reform' of Building Regulatory System Needed, Finds Dame Judith Hackitt - GOV.UK (www.gov.uk) (Accessed 3/1/2021).
5 Hackitt, J (2018) Step Up and Share. The Chemical Engineer. Step Up and Share - Features - The Chemical Engineer (Accessed 3/1/2021).
6 Hackitt, J (2018) Step Up and Share. The Chemical Engineer.
7 Hackitt, J (2017). Building a Safer Future: Interim Report. HMSO, London, UK.
8 Hackitt, J (2018) Step Up and Share. The Chemical Engineer.
9 See page 10 of Hackitt, J (2017). Building a Safer Future: Interim Report. HMSO, London, UK.
10 Hackitt, J (2018) Step Up and Share. The Chemical Engineer.
11 Hackitt, J (2018) Building a Safer Future: Independent Review of Building Regulations and Fire Safety: Final Report. Independent Review of Building Regulations and Fire Safety: Final Report - GOV.UK (www.gov.uk).
12 See page 5 of Hackitt, J (2018) Building a Safer Future: Independent Review of Building Regulations and Fire Safety: Final Report.
13 Hackitt, J (2018) Step Up and Share. The Chemical Engineer.

14 Devonshires (2018) The Hackitt Report: An Overview. Hackitt-Review.-1.pdf (devonshires.com) (Accessed 8/1/2021).

15 See page 19 of Hackitt, J (2018) Building a Safer Future: Independent Review of Building Regulations and Fire Safety: Final Report.

16 It is important to note that the single regulator is to be known as the Building Safety Regulator which has replaced the title of Joint Competent Authority.

17 Owen, J (2018) Hackitt: What It Could Mean for You. Building Hackitt: What It Could Mean for You | Features | Building (Accessed 4/2/2021).

18 See recommendation 9.1, page 109 of Hackitt, J (2018) Building a Safer Future: Independent Review of Building Regulations and Fire Safety: Final Report.

19 See page 30 of Hackitt, J (2018) Building a Safer Future: Independent Review of Building Regulations and Fire Safety: Final Report.

20 See page 45 of Hackitt, J (2018) Building a Safer Future: Independent Review of Building Regulations and Fire Safety: Final Report.

21 Barratt, L (2018) The Hackitt Review: Key Recommendations at a Glance. Inside Housing. Inside Housing - Insight - The Hackitt Review: Key Recommendations at a Glance (Accessed 4/1/2021).

22 Construction Leadership Council (2019) Raising the Bar-Improving Confidence Building a Safer Future. Raising the Bar-Improving Competence Building a Safer Future. Construction Leadership Council.

23 Gerrard, N (2018) Hackitt: 8 Key Recommendations. Construction Manager Hackitt: Eight Key Recommendations – Construction Manager (constructionmanagermagazine.com) (Accessed 9/1/2021).

24 See page 109 of Hackitt, J (2018) Building a Safer Future: Independent Review of Building Regulations and Fire Safety: Final Report.

25 Jessel, E (2018) RIBA Slams Hackitt's Post-Grenfell Review as a "Major Missed Opportunity". Architects Journal RIBA slams Hackitt's Post-Grenfell Review as 'major missed opportunity' (architectsjournal.co.uk) (Accessed 3/2/2021).

3 The Government response and the draft Building Safety Bill

Introduction

In the wake of the Grenfell Tower fire, the Government set up an independent expert advisory panel to provide advice and make recommendations to the Secretary of State for the Ministry of Housing, Communities and Local Government (MHCLG) on immediate building safety measures that should be implemented for HRRBs. On advice received from this expert panel, the Government effectively banned the use of combustible materials in the external walls and balconies of new HRRBs with effect from 21 December 2018. The actions of the expert panel are explored in greater detail in Chapter 6.

Following the publication of the final Hackitt Report, the Government accepted all its findings and all 53 recommendations of the Report whilst recognising that it would take time to build the necessary competence to enact the proposed reforms.

In December 2018, the Government published an implementation plan setting out how they proposed to action the recommendations from the Hackitt Report. The key proposals in the plan set out a stronger, more effective regulatory framework with clearer guidance on building standards and better engagement with residents, giving them a stronger voice. The end goal was to create a paradigm shift in culture across the construction industry.[1]

In June 2019, the Government published and promoted the "Building a Safer Future" consultation which set out proposals for achieving this long-term reform. A total of 871 responses to the consultation were received, representing a cross section of stakeholders from the built environment and fire safety sectors.

In addition to these initiatives, the Government put in place a number of interim measures to improve the safety of the residents:

DOI: 10.1201/9781003092803-3

- An initial £600 million of funding was made available to support building owners to remove and replace unsafe ACM cladding systems to HRRBs in both the public and the private sectors.
- In March 2020, the Government provided a budget of £1 billion pounds to support the remediation of unsafe non-ACM cladding systems to HRRBs in both the public and private sectors.
- A full technical review of Approved Document B guidance was carried out, covering the provision of sprinklers, wayfinding and evacuation alert systems in newly built flats.
- A wholesale review of the Housing Health and Safety Rating System was initiated.
- Work began with the Early Adopters Group to establish the "Building a Safer Future Charter", a new, voluntary approach, established to create an industry committed to putting people's safety first. The Considerate Constructors Scheme (CCS) has been appointed to develop and manage the Charter with the intention that the Charter will facilitate a culture change and increase public reassurance and trust in the construction industry by placing building safety above all other priorities. The Charter consists of five commitments that are a key mechanism in driving culture change and improving current practices and behaviours.[2]

The draft Building Safety Bill

Background

The results of the "Building a Safer Future" consultation were published in April 2020. At the core of the resultant proposals, which form the framework for the draft Building Safety Bill, is a new, more stringent regulatory regime for HRRBs which features[3]:

- The introduction of duty holders who will be accountable and will have statutory responsibilities for continuously managing risks across the design, construction and occupation of buildings
- Gateways (stop/go decision points) which will demand rigorous inspection of regulatory requirements at these key stages to ensure that building safety risks are considered during planning, design and construction
- A requirement for a "golden thread of information" to be created, stored and updated, using a digital platform to capture both the Gateway processes and all changes and events throughout the building's lifecycle

- The introduction of the Building Assurance Certificate which confirms that a building is fit for occupation and will be periodically reviewed throughout the life of the building
- A statutory duty on the Accountable Person, who is the duty holder in occupation, not only to assess building safety risks but also to take all reasonable steps to prevent the occurrence of a major incident within the building arising from the identified risks
- A statutory requirement to provide a "Safety Case report" which sets out how building safety risks are being identified, mitigated and managed on an ongoing basis
- A requirement for the Accountable Person to appoint a competent Building Safety Manager (BSM) to support them in managing fire and structural safety risks on a day-to-day basis
- Recognition that resident engagement and collaboration are core to keeping a building safe through improved transparency and effective handling of complaints and concerns.

The stringent regime for HRRBs is coupled with improvements and amendments to the regulatory regime that applies to all buildings including:

- The Building Safety Regulator has a duty to establish a new industry-led committee to advise on industry competence and drive continuous improvement in levels of competence.
- The Building Safety Regulator will be responsible for oversight of the competence and performance of registered building inspectors and the building control bodies in which they work.

Overview

The draft Building Safety Bill (Bill CP 264) was published on 20 July 2020. The wholesale reforms set out in this quite complex Bill, represent fundamental changes to the current regulatory system for building control and safety in England. In the foreword to the draft Bill the Secretary for State for Housing Communities and Local Government, Robert Jenrick MP states, *"at the heart of this new regime is the safety of residents in high-rise buildings"*. The aim is for the construction industry to be driven by greater accountability. Those actors who are responsible for building and managing high-rise buildings will need to evidence specifically how they considered residents safety in all work undertaken, and residents will have access to the safety information for their own building. Tougher sanctions are proposed for those responsible for managing safety risks who fail to meet their statutory obligations. The role of the powerful new Building Safety Regulator, working within the HSE, will hold responsibility for overseeing the safety and performance of all buildings and for implementing and enforcing the more rigorous regime for higher-risk buildings.

Explanatory Notes

A set of Explanatory Notes (EN) was prepared by the Ministry of Housing, Communities and Local Government and was issued alongside the draft Bill. The EN provide background information on the development of policy and explain what each part of the Bill will mean in practice. The EN are intended to be read alongside the Bill.

It is acknowledged in the EN that the draft Bill will be subject to scrutiny by a Parliamentary Committee which will provide feedback and recommendations before the Bill is finalised. Once the MHCLG has completed the stakeholder consultation and fed back the results of the consultation, the draft Bill will be introduced in the House of Commons and will take its course through the usual parliamentary process.

It is important to note that the EN state that some aspects of the proposed new regime will be taken forward using secondary legislation. For example, the competency requirements will be progressed via the Approved Documents whilst the design and construction gateway policy will be set out by incorporation within a Statutory Instrument.[4]

Definitions

In-scope buildings

The draft bill does not make any reference to "High-Rise Residential Buildings" or "High-Risk Residential Buildings" but instead adopts the term "Higher-Risk Buildings". The definition of a "Higher-Risk Building" has yet to be prescribed by the Secretary of State (s.19 of the Bill), but Clause 228 of the Explanatory Notes sets out the preferred definition of a Higher-Risk building as follows:

A building which satisfies the height condition and contains:

a   Two or more dwellings, that is, house, flat or serviced apartment
b   Two or more rooms for residential purposes, for example, supported accommodation or
c   Student accommodation

Where:

1   The height condition is that:

  a   The floor surface of the building's top storey is 18 m or more above ground level, ignoring any storey which is a roof-top plant and machinery area or any storey consisting exclusively of plant and machinery rooms; or

  b   the building contains more than six storeys, ignoring any storey which is below ground level;

  And

2   "Room for residential purposes" means a room, other than in a dwelling, which is used by one or more persons to live and sleep but excluding a room in:

  a   a residential care home;

  b   a secure residential institution, for example, prison and detention centre;

  c   temporary accommodation, for example, a hotel, hostel, guest house, hospital and hospice.

Building safety risks

Within the draft Bill, a building safety risk is defined as a risk to the safety of persons arising from a fire or due to structural failure (s.16) and a major incident is defined as one in which a building safety risk occurs and results in a significant number of deaths or serious injury to a significant number of people. Although the ill offers no definition of "significant", the moral argument must surely be that one death or one serious injury is one too many and would be deemed significant.

Structure of the Bill

There are 119 sections in the draft Bill, divided into 5 Parts and 8 Schedules. Each of the five Parts has been sub-divided into different chapters. The Parts are set out as follows:

• **Part 1** Introduction and Overview of the Act
• **Part 2** The Building Safety Regulator and how the proposed safety regime will function. Includes definitions and the scope of the regime.

- **Part 3** Amendments to the existing Building Act as it applies in England; the role of each of the duty holders; the three gateways and regulation of the building control profession including the establishment of a register of building control approvers and building inspectors.
- **Part 4** Relates to the occupation phase of the higher risk building and sets out the roles and obligations of the Accountable Person and the Building Safety Manager, including details of contraventions and sanctions.
- **Part 5** Provides supplementary and general provisions including the establishment of a New Homes Ombudsman scheme and the regulation of architects in accordance with new competency requirements.

It is envisaged that the new regulatory regime will take effect in Autumn 2021, but there is no indication of when the draft Bill will be enacted nor when the provisions will come into force.

Part 1 Introduction and Overview of the Act

Part 1 is intended to assist the reader of the Bill to understand the provisions that follow. It sets out the reason why the Bill has been brought into existence and signposts the main elements of the Bill.

Part 2 The Building Safety Regulator

Part 2 of the Bill provides for the formation of the Building Safety Regulator (BSR) as a new division within the Health and Safety Executive to underpin the core regulatory reforms in the new regime. Parts 2 and 4 of the draft Bill refer to "the regulator", which means the Health and Safety Executive. The regulator will have oversight of the building safety and performance system during all stages of the life cycle of a higher-risk building and will have three broad functions:

1   **Implementing the new more stringent regulatory regime for higher-risk buildings** – The regulator will take over the building control regime for all building work on higher-risk buildings and will be responsible for checking the work to verify compliance with the Building Regulations. This prevents developers of higher-risk buildings from choosing the building control authority. The regulator is also responsible for overseeing and enforcing the new safety regime during the occupation phase.

2 **Overseeing the safety and performance of all buildings** –
There are two key aspects to this function:

    a   Overseeing the performance of other building control bodies such as local authorities and building control approvers (currently known as Approved Inspectors). This will involve collecting and analysing data on the performance of these bodies and imposing sanctions in instances of poor performance.

    b   Advising on existing and emerging building standards and providing advice on changes to the regulations. The regulator will also commission advice on risks in, and standards of building.

3 **Assisting and encouraging competence among the built environment industry and registered building inspectors** – This function has two key workstreams:

    a   assisting and encouraging improvements in the competence of the built environment sector; and

    b   establishing a unified building control profession with common competence requirements for registration across both the private and public sectors.

To assist in carrying out its functions, the Bill gives the regulator the duty to exercise its powers under Section 11A(3) of the Health and Safety at Work, etc. Act 1974 to set up and maintain three committees:

- **Building Advising Committee** – A response to the Hackitt Review's recommendation that the Government should create a new body to oversee the built environment sector and provide expert advice, this committee replaces the Building Regulations Advisory Committee.
- **Committee on Industry Competence** – Establishing this committee addresses Hackitt's concern that the current landscape for ensuring competence in the industry is complex and fragmented and has led to inconsistency in standards. This industry-led committee will advise both the regulator and the built environment industry on industry competence.
- **Residents Panel** – The Panel will consist of residents from higher-risk buildings who will be able to contribute to key policy changes made by the BSR. The BSR must consult the Residents Panel before issuing or revising guidance which affects residents.

Plans and reports

As part of its role, the BSR is obliged to develop a strategic plan, setting out how the BSR proposes to fulfil its building functions in the period to which the plan relates (s.23). The BSR must consult with the Resident Panel prior to issuing the plan and the plan must be approved by the Secretary of State. The BSR must also issue an annual report on the information provided under the mandatory reporting requirements and the report must include a statement of how the BSR has engaged and consulted with the residents of higher-risk buildings.

Enforcement

The BSR is mandated to set up a multi-disciplinary team of authorised officers to carry out the relevant building functions. The Bill sets out the sanctions for obstructing or impersonating an authorised officer which will be a criminal offence triable in the Magistrates' Court only and will carry a fine of £1000. The offence of providing false or misleading information to the BSR is triable in the Magistrates' Court or the Crown Court. If tried at Crown Court the maximum penalty is an unlimited fine and/or two years imprisonment.

Fees and charges

The BSR will be self-funded, as recommended by the Hackitt Report, from the fees and charges it will levy on those it regulates under Parts 2 and 4 of the draft Bill. The level of the fees has still to be set in subsequent regulations.

Right of appeal and review

The decisions of the regulator can be reviewed (s.30) by people who are impacted by the decision, though secondary legislation will be required to set out which categories of decisions will be reviewable and who can apply for a review. Any reviewed decisions can be appealed to the First-tier Tribunal but only after an internal review by the regulator. This mirrors the review regime employed by the Health and Safety Executive and the Regulator of Social Housing.[5]

Clause 34 sets out a mandatory process by which the Secretary of State can periodically appoint an independent reviewer to assess the effectiveness of the building regulatory regime and make recommendations, as necessary, to ensure that the whole system, including the regulator, is fit for purpose. The Secretary of State must appoint the reviewer within five years of the Act being passed. Once the independent reviewer has completed the report, it must be published by the Secretary of State.

Part 3 Amendments to the Building Act 1984

These amendments are critical if the new regime for higher-risk buildings during the design and construction phase is to be effective and a unified professional and regulatory structure for building control is to be established. With these objectives in mind, the amendments will have the following effects:

1   The regulator will be the building control authority for all higher-risk buildings and for all work that causes any building to become a higher-risk building. This will prevent a developer from choosing the building control authority for works relating to a higher-risk building. Local authorities will be the building control authority for all other buildings.

2   The Hackitt Review recommended the introduction of new building control procedures to increase the regulatory oversight of the design and construction of higher-risk buildings. As the existing provisions within the Building Regulations are considered insufficient to implement the new stricter controls, the draft Bill allows for the provisions of Gateways and refurbishment procedures (s.37).

3   The duty holders identified in the Hackitt Review, which includes the Client, Principal Designer and the Principal Contractor appointed under CDM 2015, are to be held accountable for building safety during the construction phase and will have formal responsibilities for compliance with building regulations (s.38).

4   The existing powers within the Building Act 1984, together with amendments to the same Act set out in the draft Bill, will allow for the creation of the new Gateway regime. The Gateway regime will ensure that building safety risks must be considered at each stage of the design, construction and occupation of a building. Statutory instruments will be issued to set out the practical details of how the Gateway regime will operate.[6]

5   As part of the Mandatory Occurrence Reporting System, duty holders will need to report on any safety incidents that occur whilst building work is taking place. The list of reportable occurrences will be provided by secondary legislation.

6   The current building regulations do not contain any specific provisions about the competence of people undertaking building work. New provisions are included in the draft Bill to impose a general duty of competence on people carrying out work on all buildings to ensure compliance with the building regulations (s.39).

7   Under the current regime, developers can choose to build at risk without receiving building control approval. This will become an offence under the new regime for higher-risk buildings. Applicants may appeal to the Secretary of State for a decision in cases where the regulator has failed to make a decision within the prescribed timescale.

8   Where work is carried out but contravenes the building regulations, the time limit for enforcement action is set by sections 35 and 36 of the Building Act 1984. An amendment to the Act extends the time limit to take out prosecutions for contravention of the building regulations from two to ten years and the requirement to correct previously completed non-compliant work from one to ten years.

9   New powers are also to be introduced to enable building control authorities to address non-compliance with building regulations without having to take out a criminal prosecution. More importantly, building control authorities will be able to issue stop notices to prevent any further work being done on an already non-compliant building. These measures replicate the issuing of notices under Section 21 of the Health and Safety at Work, etc. Act 1974.

10 If a corporate body commits a criminal offence under the Building Act, which covers the duty holder roles and requirements for inspection and reporting on buildings, any director, manager, secretary or similar officer of that body is also deemed to have committed that offence if the individual has consented to or connived in the commission of the offence or if the offence is attributable to their neglect.

11 The Bill introduces new work titles. Individuals and organisations currently known as "Approved Inspectors" will need to register as "Building Control Approvers", that is, private sector building control. The registered building inspector is a new role, providing advice to local authorities or registered building control approvers overseeing building work. Many inspectors in local authorities and Approved Inspectors are expected to transition to this role. The Regulator will establish and maintain a register of building inspectors (public sector) and building control approvers (private sector) as well as a code of conduct for building inspectors and a set of

professional conduct rules for registered building control approvers (Section 44).

12 The Regulator must investigate claims of misconduct and, if the case is proven, make a disciplinary order, which will include suspension of an individual's registration. If someone impersonates a registered building inspector or a registered building control approver, it is an offence punishable by a fine.[7]

13 The Regulator must keep a register of specified relevant information provided to building control. This will be a national electronic register or portal. It will contain notices, certificates, orders, consents, demands and plans. This must be maintained electronically and certain parts of it will available for inspection by members of the public (see Section 52).

Part 4 Higher-risk buildings

The provisions in Part 4 concern the occupation phase of higher-risk buildings. It sets out a new general framework for how these buildings will be managed and the responsibilities assigned to the various duty holders. The Hackitt Report outlined the idea that the "golden thread" of building and fire safety information should be captured digitally during the design and construction phase and subsequently carried over into the occupation and operational management phase of the building's life cycle. During the occupation phase, statutory duties will be imposed on the accountable person and the building safety manager to ensure that all risks to building safety are flagged up and addressed in advance thus reducing the risk to residents and users of higher-risk buildings.[8]

The Accountable Person

The Accountable Person in relation to a higher-risk building is defined as follows:

a a person who holds a legal estate in possession in any part of the common parts, or

b a person who is under a relevant repairing obligation in relation to any part of the common parts.

The definition of "the common parts" is set out in S.61(3) to mean "*the structure and exterior of the building*" or "*any part of the building provided for the use, benefit and enjoyment of the residents of more than one dwelling*".

The Government acknowledges that building ownership in the UK is complex and the definition has been drawn up to ensure that the correct person is defined as the Accountable Person and will be responsible for the statutory obligations under the building safety regime. This will usually be the Freeholder of the property, but it may also include the head lessees, management companies, housing associations, right to manage companies or a long leaseholder (with a lease in excess of 21 years).[9] Interestingly, neither the draft bill nor the EN specify whether the Accountable Person will be a named board-level individual as stated in the Government's 2019 Consultation document but perhaps this requirement has now been negated.[10]

Duties

The Accountable person has a number of statutory obligations which include:

- Registering the building with the Regulator (s.62)
- Applying for a Building Assurance Certificate

This is a new term for the document previously referred to as the Building Safety Certificate. The Regulator must issue this certificate if it is satisfied that the AP has not contravened any of the relevant duties. The AP must display the Building Assurance Certificate in a conspicuous position within the subject building.

- Appointing a Building Safety Manager:
  Before the building is occupied, a Building Safety Manager must be appointed with the appropriate competence (s.67) (i.e. skills, knowledge, and experience) to identify and reduce building safety risks as soon as possible (s.72) to avert any major incident (s.73). Further assessments must be carried out at regular intervals.
  If the AP fails to appoint the BSM, they have committed an offence which is punishable by imprisonment and/or a fine. If the AP appoints a corporate body as the BSM, they in turn must appoint a "nominated individual" to carry out the functions of the BSM. The regulator may veto an appointment of a BSM if the person is not suitable.
- Preparing a building "safety case report":
  The AP must prepare a building "safety case report" which assesses the building safety risk (s.74) and frames how the BSM is to manage the building (s.76). The BSM must notify the AP if any assessment

is no longer valid and if further action should be taken. The safety case report will include a full building description, a hazard and risk assessment, a summary of mitigation measures and the approach taken to risk management. Compiling the report might, depending on the complexity of the building, require contracting a team of technical experts such as structural engineers, fire engineers and safety experts.

• Producing a residents' engagement strategy that promotes "the participation of relevant persons in the making of building safety decisions" (s.82)
• Providing prescribed information to the regulator and residents (s.80):
    It is reasonable to assume that this information will equate to the "golden thread" of building information described in the Hackitt Review.
• Establish a system for investigating complaints (s.84)
• An outgoing AP must pass the prescribed information onto the new or interim AP (s.81).

Residents[11]

The Hackitt Review emphasised that residents need to be placed at the centre of the new regime and that they need to have a legitimate voice and access to the correct and current information. The draft Bill sets out detailed arrangements to promote the participation of residents and flat-owners in decision-making around building safety risks within their own building. The AP is required to produce a Resident Engagement Strategy (RES) which will presumably either be used as part of the Gateway 3 requirements or will be included within the Safety Case prepared for the occupation phase of a higher-risk building.

The Bill sets clear criteria for what the RES must include. It details the information that the AP must provide to residents concerning the management of the building, the scope of the issues that the residents must be consulted about and how they are to be consulted. It is anticipated that the Government, in conjunction with the Residents Panel, will draw up and publish a good practice guide to assist APs in developing a suitable and effective RES. In addition to the information provided within the RES, residents will be able to request more detailed safety information such as fire risk assessments, planned maintenance and repair schedules, building safety inspection reports, the fire strategy

for the building, information on the maintenance of existing fire protection measures and details of all planned and historical changes to the
buildings.

The AP will be required to set up a complaints procedure to deal
with residents' safety concerns and the BSM will have to adhere to
the published procedure when handling complaints. The Regulator is also required to establish a complaints procedure to deal with
complaints that are escalated to them by a resident or via the AP or
BSM. The detail of both complaints and procedures, with required
response times, will be set out in secondary legislation.

Residents duties

Under the Bill, residents are subject to a set of express obligations to
keep their properties in good repair and working order. This includes
all relevant residents' items such as electrical appliances or gas safety in-
stallations. They are obliged to take reasonable care to avoid damaging
safety items in the common parts and to comply with any request from
the AP for information in connection with the AP's duties relating to
building safety risks. If an AP believes a resident has breached any of
their statutory obligations, then the AP is able to serve a notice on a
resident requiring them to provide certain information or stop a certain
behaviour. This can, if necessary, be enforced by an order from the
County Court. It is interesting that while the Hackitt Report stated
that residents should co-operate with the AP, the Bill only refers only
to these specific duties as opposed to any general duty to co-operate.

The AP will have rights to enter a residents dwelling following an
application to the court. Access may be needed by the AP to determine
whether a resident is adequately discharging their obligations to assess
and minimise any building safety risks. However, there does appear to
be an anomaly in that residents have to avoid damaging safety items in
the common parts but not in their own flat. This could, potentially, be
an issue as it is not always clear within lease terms as to whether a flat
entry door is demised to the residents.[12]

Building Safety Charges

The scope of the Building Safety Charge covers the cost of all the
building safety measures carried out, including overheads. This can include appointing the BSM, drawing up the safety case document, the
costs of any building safety works, compliance costs, etc. The IA sets

out budget costs for these works, which demonstrates how expensive they may be. The IA assumes that where buildings currently have no plans or inaccurate plans, it will cost between £10,000 and £19,000 per building to produce a two-dimensional Computer Aided Design (CAD) plan and evaluation drawing. Other options such as 3D scanning and photogrammetry will be more expensive.[13] It is estimated that the appointment of a BSM may cost £3,000 per building,[14] and if the BSM is to undertake all their designated duties, this could cost £6,400 to £10,000 per building per annum.[15]

In terms of who pays these costs, the proposed legislation will amend the Landlord & Tenant Act 1985 so that every tenant with a long lease must pay the Building Safety Charge within 28 days of demand. A "long lease" is defined as a lease granted for a term of over 21 years, including Right to Buy and Right to Acquire leases and also shared ownership leases. The new sections to be inserted in the Act mirror the existing statutory consultation regulations so that long leaseholders will be consulted, though this can be dispensed with if the works are deemed to be urgent. The cost cannot be recovered if the long leaseholders do not receive a demand or a notification of costs within 18 months of the cots being incurred. Leaseholders are, potentially, facing massive bills which the IA has estimated to be £78,000 per leaseholder.[16] Leaseholders may have to bear the cost of putting right the faults of others such as, developers, contractors or those producing the Government's own advisory documents.[17]

Part 5 Supplementary and general[18]

The 'New Homes Ombudsman'

The Secretary of State is to establish a scheme whereby owners and occupiers of new-build homes can make complaints to the newly created role of New Homes Ombudsman. Residential developers can also be members of the scheme. It is anticipated that the Secretary of State will issue or sanction a code of practice for standards of conduct and quality of work (s.106).

Housing Ombudsman

Social housing complainants will be able to escalate a complaint directly to the Housing Ombudsman, though they will have to complete their landlord's in-house complaints procedure first. This will put social housing tenants on parity with those of other tenures.

Construction products

Schedule 8 of the draft Bill allows the Secretary of State to regulate all construction products that are made available in the UK. The Regulatory regime will identify "designated products" and issue a statutory list of "safety critical" products, where the failure of a product would result in death or serious injury. All other construction products that are not designated or deemed to be safety critical will be subject to a general safety requirement.

Changes to registration of architects

Section 111 of the draft Bill amends the Architects Act 1997 to allow the new Architects Registration Board to monitor the competence of architects throughout their registration. If an architect does not meet the requirements deemed necessary by the Board, the architect may be removed from the register of architects. Any disciplinary orders, made by the Professional Conduct Committee of the Architects Registration Board, will be listed alongside an architect's entry in the Register, to allow for greater transparency for fellow professionals and for public alike.

Fire safety reform

The Government completed a wholesale review and reform of the management of fire safety in buildings by introducing the Fire Safety Bill 2020. The Bill complements the change in regime set out in the draft Building Safety Bill.

The Fire Safety Bill

The Fire Safety Bill was introduced in 2020 and amends the Regulatory Reform (Fire Safety) Order 2005 with far-reaching consequences. External walls and anything attached to those walls – that is, cladding – and flat entrance doors of multi-occupied residential buildings, of any height, fall within the scope of that Order. Crucially, the Government's position is that the Fire Safety Bill is a "clarification" of the Fire Safety Order, rather than an amendment. This is contentious, as the inclusion of external walls within the Bill is not merely a clarification, but actually extends the scope of the Fire Safety Order.

The retrospective effect of the "clarification" has far-reaching implications for duty holders as it means that external walls will be deemed to have always fallen within the scope of the Fire Safety Order. This could have a knock-on effect with respect to litigation over liability for the replacement of external cladding systems which were installed pre-Grenfell and signed off as compliant with the Building Regulations in force at that time, but which fire authorities may now allege were always in breach of the Fire Safety Order, according to the "clarification" set out in the Fire Safety Bill.[19]

Summary

The implementation of these building safety reforms should drive the move away from the fragmented system that allowed the fire at Grenfell Tower to occur and towards a system where a new BSR has a holistic oversight of building safety on higher-risk buildings. By improving the system for inspecting the quality of building work and identifying key duty holders who will be responsible for the safety of residents throughout the life cycle of higher-risk buildings, these reforms, including the creation of the Residents Panel, should meet the objective of making residents feel safe and be safe in their homes.

There have been criticisms of the draft Bill. The height of a building is considered by some to be a poor measure of risk to human life and believe the Government should have considered other factors when determining risk. It is accepted that it is logistically impractical to include the entire building stock in England under the remit of the new Bill, but it has been suggested that if the scope of the legislation is widened then the next consideration should include buildings where vulnerable people reside.

Many buildings have more than one owner, leading to the possibility that a building may have multiple APs. The complexity of such a situation could, potentially, cause uncertainty and disputes. To mitigate this, there should be a general duty written into the Bill for APs in these circumstances to co-operate with each other.

The role of the BSM will be a core part of the new regime but the skill-set required for this role is so specialised that the industry will need to train people to become BSMs prior to the enactment of the legislation. If the new safety regime is to be implemented quickly and well, it has been recommended that the final Bill should include both the new competence framework and the accreditation and registration system for the BSM role.[20]

Notes

NB: The main documents used in providing the commentary to the draft Building Safety Bill in this chapter are The Draft Bill, the Explanatory Notes that accompany the Bill and Towers & Hamlin Essential Guide to the Draft Building Safety Bill: Ministry of Housing, Communities and Local Government (2020) Draft Building Safety Bill Draft Building Safety Bill – GOV.UK (www.gov.uk) (Last Accessed 19/2/2021).

1 Graye, L (2020) How Will the New Building Regulatory Framework for Higher Risk Residential Buildings Materialise in the UK. Unpublished thesis. London South Bank University.
2 Building a Safer Future (2020) Building Safety Starts with You. Building a Safer Future – Building Safety Starts with You (Accessed 12/1/2021).
3 See pages 7 and 8 of the Explanatory Notes to the Draft Building Safety Bill: Ministry of Housing, Communities and Local Government (2020) Draft Building Safety Bill: Explanatory Notes. Draft_Building_Safety_Bill_PART_2.pdf (publishing.service.gov.uk) (Last Accessed 19/2/2021).
4 Trowers & Hamlin (2020) Draft Building Safety Bill: Essential Guide. Essential guide to the draft Building Safety Bill - Trowers & Hamlins (Accessed 14/1/2021).
5 Trowers & Hamlin (2020) Draft Building Safety Bill: Essential Guide.
6 See pages 11 and 12 of the Explanatory Notes to the Draft Building Safety Bill.
7 Trowers & Hamlin (2020) Draft Building Safety Bill: Essential Guide.
8 Bright S (2020) The Draft Building Safety Bill and Higher-Risk Buildings: Overview and Definitions. The draft Building Safety Bill and Higher-risk buildings: Overview and Definitions | Oxford Law Faculty (Accessed 17/1/2021).
9 Bright S (2020) The Draft Building Safety Bill and Higher-Risk Buildings: Overview and Definitions.
10 Trowers & Hamlin (2020) Draft Building Safety Bill: Essential Guide.
11 The commentary on the residents' section in the draft Building Safety Bill relies heavily on the Trowers & Hamlin Essential Guide.
12 Bright S (2020) The Draft Building Safety Bill and Higher-Risk Buildings: Key Duties. The draft Building Safety Bill and Higher-risk buildings: Key Duties | Oxford Law Faculty (Accessed 18/1/2021).
13 See Clause 241 of the Impact Assessment to the Draft Building Safety Bill.
14 See Clause 321 of the Impact Assessment to the Draft Building Safety Bill.
15 Bright S (2020) The Draft Building Safety Bill and Higher-Risk Buildings: Costs and Complex Buildings. The draft Building Safety Bill: Costs, and complex buildings | Oxford Law Faculty (Accessed 18/1/2021).
16 See page 64 Table 34 of the Impact Assessment to the Draft Building Safety Bill.
17 Bright, S (2020) The Draft Building Safety Bill and Higher-Risk Buildings: Costs and Complex Buildings.
18 The commentary on Part 5 of the draft Building Safety Bill relies heavily on the Trowers & Hamlin Essential Guide.

19 Crown Office Chambers (2020) A New Dawn in Building Safety? A new dawn for Building Safety? – Crown Office Chambers (Accessed 18/1/2020).
20 House of Commons Committees (2021) Does the Draft Building Safety Bill improve Building Safety? HC&LG Select Committee Does the draft Building Safety Bill improve building safety? (shorthandstories.com) (Accessed 19/2/2021).

4 Competency frameworks

Introduction

The Hackitt Review made it clear that one major flaw in the current regulatory system is the lack of consistency in the processes and standards for assessing the skills, knowledge and behaviours of those working on HRRBs. The Review highlighted the fragmented approach to assessing competency and the lack of relevant qualifications. The different approaches to competence standards and assessment have led to actors in the industry concentrating on their individual specialisms rather than viewing their work holistically and treating a building or a development as a single system. This approach has resulted in responsibility being widespread with no single person or organisation accepting the primary responsibility for safety throughout the life of a building, and it has jeopardised the fire safety of HRRBs.

In order to address these failings, the CSG for "Building a Safer Future" was set up by the MHCLG to take forward the recommendations on competence put forward by the Hackitt Review. The membership of the CSG included professionals from the construction industry, the fire safety sector, building owners and managers. The CSG was tasked with developing the role and remit for an overarching competence body which would:

• ensure a coherent and consistent approach to raising and overseeing competence standards within each discipline in scope; and
• support the delivery of competent people working on HRRBs.

Twelve working groups were formed to develop competence frameworks for the individual actors. These groups were Engineers (WG1), Installers (WG2), Fire engineers (WG3), Fire risk assessors (WG4), Fire safety enforcing officers (WG5), Building standards professionals (WG6), Building designers, including architects (WG7), the new role of Building safety managers (WG8), Site supervisors (WG9), Project managers

DOI: 10.1201/9781003092803-4

(WG10), Procurement professionals (WG11) and Products professionals (WG12). An additional separate working group, WG0, was set up to define the role of the overarching body and recommend how the system for overseeing competence should operate to provide assurance that all actors involved in the design, construction, inspection, maintenance and management of HRRBs are fully competent to carry out their roles. In total, more than 150 institutions and 300 professionals collaborated to produce the interim report "Raising the Bar" (2019).[1]

Raising the bar

The interim 600-page report, released in August 2019, proposed nominating existing trade bodies and professional institutions to over-see the training and accreditation within the individual trades and professions. In certain instances, such as installers and the new role of building safety manager, a new accreditation body needed to be established. It was envisaged that each organisation awarded an accreditation responsibility would in turn have to accredited by a qualified body. In addition, the overarching body, provisionally named "the Building Safety Competence Committee", would set the standards for the competence frameworks and issue guidance notes to drive ongoing improvement. Adoption of the new framework would be driven by the draft building safety bill, as only those verified under the competence framework would be qualified to take on the new roles of principal designer, principal contractor and building safety manager.

It was acknowledged in the interim report that the complexity of the vision it set out would provide challenges in rolling it out across the industry. If the frameworks became too bureaucratic, it may deter professionals from "buying into" the new vision. It was also recognised that a greater degree of independent, regular scrutiny was required, and following the consultation on the interim report, it was proposed that existing arrangements for third-party assessment should be improved by requiring that all assessments should include the necessary competencies for working on HRRBs.

Setting the bar

Following a period of extensive and widespread consultation, the final report *Setting the Bar: A New Competence Regime for Building a Safer Future* was published in October 2020. The aim of the CSG in producing this final report was *"to come up with a blueprint to improve competence for those working on higher-risk buildings and drive a culture change right across the industry*".[2] The final report recommendations lay the foundations for achieving a coherent and consistent approach to assessing and ensuring competence across the sectors. It is anticipated that, working in tandem with the legislative changes set out in the draft Building Safety Bill, this approach will drive a far-reaching culture change in which all actors within the industry recognise and embrace their responsibility for delivering safe buildings.

The new regime puts in place a comprehensive framework of competence standards, developed by the various working groups on an individual sector basis and supported by third-party assessment for individuals and companies, with corresponding third-party accreditation of those responsible for conducting the assessments. The proposed system of competence comprises four main elements[3]:

1   A new competence committee sitting within the BSR;
2   A national suite of competence standards;
3   Arrangements for independent assessment and reassessment against the competence standards; and
4   A mechanism to ensure that those assessing and certifying against the standards have appropriate levels of oversight.

The national suite of competence standards (element 2) will encompass[4]:

- A British Standard for an overarching competence framework that will be common across all disciplines in the industry
- Public Approved Specification (PAS) standards for the three regulated roles of Principal Contractor, Principal Designer and Building Safety Manager
- A series of sectorial competence standards or frameworks developed by the 12 working groups. These will set out discipline-specific requirements for the skills, knowledge, experience and behaviours needed to carry out the specific roles, with a rigorous approach to training and assessment.

Some of the proposals have already been acted upon as the draft Building Safety Bill states that one of the key functions of the BSR is to assist and encourage competence in the built environment. Likewise, the MH-CLG has tasked the British Standards Institute to develop and produce the National Standards for the overarching competence framework and the PAS standards for the three regulated roles of Principal Designer, Principal Contractor and Building Safety Manager. It is anticipated that the finalised PAS documents will be published by March 2022.[5]

Third-party assessment of individuals and organisations

The principal objective is that every person working on a higher-risk building must be subject to a system of competence assessment and management to make certain that they are competent to deliver safe outcomes. The assessments will be undertaken by independent third-party organisations such as professional or licensed bodies which are separated by "clear blue water" from the work that the person undertakes. However, a pragmatic approach has been taken, recognising that in some circumstances, it will not be feasible to require every person to be third party assessed. In these instances, any non-assessed person must be properly supervised. It is for individual sectorial competence frameworks to clearly state which roles, and at what level, supervision rather than third-party assessment is appropriate and acceptable. For example, new graduates or entrants on training schemes will need time to gain the necessary experience to become competent. It is anticipated that these arrangements will need to be approved by the building safety competence committee established within the BSR. This should ensure that all safety-critical decisions are only made by competent, third-party assessed persons.

In certain instances, companies as well as individuals will need to be third-party assessed and this service can be provided by the organisation being subjected to third-party accreditation. It will be for the Working Groups and the building safety competence committee to decide how this will operate practically in each sector.

The assessors themselves, be they individuals or organisations, must also be subject to rigorous oversight to ensure the maintenance of their standards. This complies with the point made in the Hackitt Review that a system should be put in place to *"accredit the accreditors".*[6]

Continuing professional development

Continuing professional development (CPD) requirements vary considerably within each sector, and so it is proposed that a set of common principles should be established to provide guidance to the individual sectors and to which the BSCC would refer when holding the sectors accountable.

Barriers to implementation

The Competence Steering Group recognises that by developing and publishing the draft Building Safety Bill, the Government has *"set out the biggest reform of building safety in 40 years"*[7] and that if all those actors within safety-critical professions and trades attain the higher levels of competence set down within the sector-specific frameworks, this should enable the industry to rebuild the trust with the residents and users of higher-risk buildings. However, the CSG has expressed their disappointment that the draft Building Safety Bill does not mandate those working on higher-risk buildings to be assessed against the competence frameworks. Instead, it is expected that responsible firms will want to adopt the competence frameworks. But in the current uncertain financial climate, it is reasonable to assume that without the necessary regulatory pressure, contractors and other firms who choose not to invest in complying with the competence frameworks, will be able to undercut their competitors. Thus, the culture of winning work by "gaming the system" will be perpetuated. For the adoption of the frameworks to be widespread, it is crucial that the Government takes a forceful lead in promoting and signposting the importance of using only accredited individuals and firms to drive the necessary culture change.

Safer people, safer homes: building safety management

The draft Building Safety Bill requires that all higher-risk buildings are managed by the newly created role of BSM. The BSM will be key in supporting the Accountable Person (AP) by advising the AP of any work that has taken place in the building, ensuring that the building meets with the requirements of the BSR and, crucially, communicating this information to the building's residents so they are safe and feel safe in the place where they live. WG8 was tasked with formulating and defining the competencies required for the newly designated role of BSM and their final report *Safer People, Safer Homes: Building Safety Management*[8] sets out a comprehensive framework of competencies and advises how the "Named Individual Building Safety Manager" should fit within the wider organisational picture.

The Building Safety Manager

The role of BSM has been developed around Recommendation 3.1 in the Hackitt Review which advised that[9]

> The dutyholder must nominate a named 'building safety manager' with relevant skills, knowledge and expertise to be responsible for the day-to-day management of the building and act as a point of contact for residents.

Hackitt also provided further detail on the role in Recommendation 5.4 which stated that[10]

> Relevant parties should work together, along with the relevant professional bodies, to develop and define a robust, comprehensive, and coherent system for:
>
> a   the competence requirements for the role of building safety manager of HRRBs; and
> b   the remit of this role in introducing and overseeing the process by which residents in HRRBs would be able to access fire safety awareness training.

To this end, WG8 was tasked with determining the competencies for the new BSM role and their objective was to deliver holistic life safety in whole buildings. WG8 noted the existing wide range of ownership models which can make residential property management complex. The different ownership models such as Right to Manage, Residential Management Companies and Commonhold, combine with a similar variety of different leases such as a 999-year fully repairing and insuring lease, a six-month Assured Shorthold Tenancy. This could give rise to a hierarchy of organisation, BSMs and Named Individual BSMs with responsibilities for differing areas of a single building. It is vital that these BSMs work together in a spirit of collaboration when completing a Safety Case review or determining the future roles and responsibilities at the point of registering a building at Gateway 3. It may also be necessary for the BSMs to co-ordinate their roles and duties with the Responsible Person (RP) as defined in the Regulatory Reform (Fire Safety) Order 2005, not least because the RP also has a statutory duty to liaise with the occupiers of a building.

The final report acknowledges that a Named Individual BSM may need to gain the necessary knowledge and experience to fulfil the role, and the BSM framework sets out levels of competence which relate to the complexity of the buildings being managed. The framework includes the provision for a senior Named Individual BSM to have overall responsibility for a very large or complex building whilst a less experienced BSM could act in a subordinate role, to obtain the necessary experience. It is expected that each Named Individual BSM will hold a card that states their attained level of competence with the industry body, perhaps the BSR, holding a centralised register of accredited competent BSMs.

In the final report, WG8 set out eight recommendations with respect to the BSM competence framework[11]:

1 The BSM role should:

    a   Be a role, with statutory duties and functions, responsible for life safety in whole buildings and for engaging with residents/ occupiers

    b   Sit within a wider organisational structure, the organisational BSM, so that the Named Individual BSM role is underpinned by adequate resources and support

    c   Be appointed by the AP who is a duty-holder. The AP cannot delegate their responsibilities to the BSM.

2 The Competence framework for the Named Individual BSM will cover the core knowledge, skills, experience and behaviours required for the role to be adopted for higher-risk buildings:

    a   This framework will be aligned with the overarching benchmark competence framework for higher-risk buildings.

    b   To comply with the framework, a person must demonstrate the appropriate experience in managing building risk, hold a relevant recognised professional qualification and pass a set of assessments which will test their knowledge, experience and behaviours.

    c   The Named Individual BSM should be reassessed for competence every three years when they will need to show evidence of participation on a refresher course, undertaking CPD and adherence to the Code of Conduct.

3 WG8 recommends a statutory certification and registration structure for higher-risk buildings which will include:

    a   Building Registration

    b   Registration of the BSM organisation

    c   Registration for the AP/duty holder who can be held responsible and accountable for building safety and resident engagement

    d   Third-party accredited certification for the Named Individual BSM

    e   The BSR should maintain a national register for the statutory roles of AP, BSM and Named Individual BSM.

    f   The Building Safety Competence Committee will be responsible for setting, maintaining, assessing and delivering competence standards.

4 Recommendations relating to the "golden thread" of information:

a   The content of the Safety Case file and the Fire & Emergency File should be mandated with the information managed by a competent person.

b   The information should be held digitally on a National Database.

c   The Fire & Emergency File should become mandatory for most residential buildings, both new and existing.

5 The BSM role will include responsibility for producing and implementing the resident engagement strategy for and on behalf of the duty-holder AP.

6 Safety education of the residents/occupiers should be underpinned by an extended "Fire Kills" campaign.

7 Leases, tenancy agreements and licences should be amended to include a provision to allow the right of reasonable and proportionate access to individual residential units.

8 The WG8 endorses and supports legislative changes that oblige occupiers of higher-risk buildings to behave in a safe manner and require a licence for alterations (or similar) when making structural changes to their demised property.

The scope of the draft building safety bill is limited to fire and structural safety, but WG8 believes that the regime should go further and recommends a holistic "whole systems and life safety approach". To drive this, WG8 has urged the industry to develop "whole systems" guidance, even though there will be no statutory obligation to do so.[12]

Summary

Work undertaken in the last two years and the publication of the competency reports show a determination within the construction industry to embed quality assurance through a programme of continuous third-party, impartial assessment. There is a clear intent to provide reassurance and rebuild the trust of residents by demanding professional competence in those roles that are crucial in the devising of robust safety systems to protect residents.

Accredited third-party certification of organisations together with the verification of individual competencies allows insurance and consumer protection requirements to be made mandatory. WG2 expressed the hope that the new competency requirements would operate in tandem with established systems such as the Competent Person Scheme and TrustMark to avoid firms incurring unnecessary, additional costs.[13]

General criticisms have been levelled at competency frameworks, suggesting that they focus on the past and cannot keep pace with a rapidly changing commercial landscape. There are concerns that they may be bureaucratic and not user-friendly, creating "clones" as everybody is expected to behave in the same fashion.[14] At this stage, the main barrier to effective implementation is that the draft Building Safety Bill does not mandate those working on higher-risk buildings to be assessed against the competence frameworks. Instead, it is expected that responsible firms will want to adopt the competence frameworks. However, if no regulatory pressure is brought to bear, there is a real risk that the culture of winning work by "gaming the system" will be perpetuated.

The new role of BSM clearly requires a broad breadth of knowledge alongside a detailed technical skill-set. Whilst knowledge can be measured by gaining a professional qualification, the real concerns within the industry focus on a BSM's ability to communicate with the residents of a high-risk building. This concern is driven by the fact that there is no straightforward solution to producing a resident strategy, not least because the demographic of any group of residents is invariably diverse, varied and complex. Softer skills such as effective and empathetic communication will need to be nurtured by practice with appropriate supervision for the inexperienced, aspiring BSM before they can be formally assessed.[15]

Notes

1 Competence Steering Group (2019) Raising the Bar: Interim Report. CIC. raising-the-barinterimfinal-1.pdf (cic.org.uk) (Accessed 29/1/2021).
2 See page 6 of Competence Steering Group (2020) Setting the Bar: A New Competence Regime for Building a Safer Future. CIC. setting-the-bar-exec-summary-final.pdf (cic.org.uk) (Accessed 30/1/2021).
3 See page 7 of Competence Steering Group (2020) Setting the Bar: A New Competence Regime for Building a Safer Future. CIC.
4 See page 8 of Competence Steering Group (2020) Setting the Bar: A New Competence Regime for Building a Safer Future. CIC.
5 See page 10 of Competence Steering Group (2020) Setting the Bar: A New Competence Regime for Building a Safer Future. CIC.
6 Hackitt, J (2018) Building a Safer Future: Independent Review of Building Regulations and Fire Safety: Final Report. Independent Review of Building Regulations and Fire Safety: final report - GOV.UK (www.gov.uk).
7 See page 15 of Competence Steering Group (2020) Setting the Bar: A New Competence Regime for Building a Safer Future. CIC.
8 Working Group 8 (2020) Safer People, Safer Homes: Building Safety Management. annex-8a-safer-people-safer-homes-building-safety-management.pdf (cic.org.uk) (Accessed 10/2/2021).
9 See page 53 of Hackitt, J (2018) Building a Safer Future: Independent Review of Building Regulations and Fire Safety: Final Report.

10 See page 80 of Hackitt, J (2018) Building a Safer Future: Independent Review of Building Regulations and Fire Safety: Final Report.
11 See page 10 of Working Group 8 (2020) Safer People, Safer Homes: Building Safety Management.
12 See page 14 of Working Group 8 (2020), Safer People, Safer Homes: Building Safety Management.
13 National Association of Professional Inspectors and Testers (2020) NAPIT Supportive of Setting the Bar Report. NAPIT supportive of Setting the Bar report – Electrical Contracting News (ECN) (Accessed 19/2/2021).
14 The Chartered institute of Personnel and Development (2021) Competence and Competency Frameworks Competence & Competency Frameworks | Factsheets | CIPD (Accessed 19/2/2021).
15 Newcomb, J (2020) The Role of the Building Safety Manager: What Do Landlords Need to Do When Recruiting. Inside Housing – Insight – The role of the building safety manager: what do landlords need to consider when recruiting? (Accessed 19/2/2021).

5 The golden thread and traceability

Background

The Hackitt Review highlighted significant concerns surrounding the ineffective operation of the current system of creation, maintenance and handover of building and safety information. The review also identified that where building information is handed over, it is often deficient or held in an analogue format that is not easily accessible to actors who need to access it.[1] The fragmented nature of building information creates a number of challenges. It can be unclear whether any changes have been made between the original design drawings and the "as-built" drawings handed over on completion of the building which may have a negative impact on the safety strategy of the building. The building owner may not hold the necessary information to manage building safety effectively throughout the building's life cycle nor have the information needed to refurbish the building making it difficult to ascertain what effects any changes may have on building safety. Similarly, access to accurate, up-to-date information about a building is crucial when undertaking a fire risk assessment of a building and determining whether any remedial action is required.[2]

To counter this situation, the review recommends a clear model of risk ownership with transparency of information and an audit trail readily available throughout the life cycle of a building to provide reassurance and evidence that a building was built safely and continues to be safe in occupation. An important component of this audit trail will be the creation of a "golden thread of information" specific to each HRRB. The concept of a "golden thread" is used in many business disciplines and is a commonly implemented framework for performance management to ensure that an organisation's goals, vision and values inform, and are informed by, its processes, systems and people.[3]

DOI: 10.1201/9781003092803-5

The draft Building Safety Bill includes provisions that will assist the creation of a golden thread of information and the intention of the bill's clauses is to ensure that the right people have the right information at the right time to determine that buildings are safe and to manage building safety risks effectively throughout a building's life cycle.[4] The exact components of 'the golden thread' of building information are still to be agreed upon and the draft building safety bill states that the specific information, data and documents, that make up this required information will be set out in regulations.[5]

However, the Hackitt Review is clear that the core components of the golden thread will be information on key building safety requirements such as fire safety and structural safety. To this end, Hackitt recommends that the minimum components of the golden thread will be four key information products[6] that are crucial to providing duty holders with a fuller oversight of building safety and other building regulation requirements, throughout the procurement, design, and construction of a building. The four key information products are as follows:

1   A digital record of essential building information to be used by the duty holders to demonstrate to the BSR the safety of a building throughout its life cycle. This will include the initial design intent throughout construction, and any changes that are made throughout the occupation of the building. The digital record must also provide product traceability.[7]
2   A FEF, which is central to setting out the critical fire safety information for a building.[8]
3   Full Plans – a complete set of accurate "as built" drawings, showing any and all design changes from the original tendered design drawings.
4   A Construction Control Plan, describing how building safety and Building Regulations compliance will be maintained during the construction phase and how any changes to the Full Plans, signed off at Gateway Point 2, will be controlled and recorded.[9]

The golden thread of information

The digital record

The stated purpose of the digital record is to ensure that accurate information about a building is accessible at all points throughout its life cycle. It should be a live, dynamic record of the design, construction, and occupation of new HRRBs, which must include any changes or subsequent refurbishments that may occur while the building is in occupation. It is intended that the digital record *"will inform the duty-holder about which gateway points and safety case processes apply, throughout the lifecycle of the building"*.[10] The digital records are to be held in a format which is both open and non-proprietary and includes the implementation of proportionate security controls. Clearly, the new digital record-keeping system will take some time to implement for existing buildings, as up-to-date information may not be readily available and may have to be created. However, it is expected that all new higher-risk building projects should be able to implement this record-keeping practice immediately, during the design phase. It will undoubtably require strong leadership and effective collaboration to ensure that the new digital record requirement is successfully implemented.

Dame Hackitt envisages that the scope of information in the digital record for new buildings will include

> the size and height of the building, full material and manufacturer product information, identification of all safety critical layers of protection, design intent and construction methodology, digital data capture of completed buildings e.g. laser scanning, escape and fire compartmentation information and records of inspections/reviews/consultations.[11]

To ensure that a safety case file is completed satisfactorily, the golden thread must also contain *"information on the building management system in relation to fire and structural safety, records of maintenance, inspection and testing undertaken on the structure and services and evidence that the competence of those undertaking work on the building was sufficient"*.[12]

The expectation for existing buildings is lower to avoid unreasonable pressure on existing building owners who may not have received the necessary information from the developer or from a previous owner. However, intrusive surveys may be required to provide accurate information to support the safety case. The information that should be recorded, available and maintained for existing buildings includes the size and height of the building, the type of structure and fabric of the building, escape and fire compartmentation information, systems in operation and permanent fixtures and fittings. It is for duty holders to identify any gaps in the records currently held and plan a strategy for completing and updating the information.

The Fire Emergency File

In addition to the digital record, Hackitt recommends that the golden thread should also include an FEF.[13] The FEF has been identified as core to setting out crucial fire safety information for a building. The FEF will be created by the client and updated by the Principal Designer and Principal Contractor during the design and construction phase, before being handed over to the AP, in order to enable the AP to understand how best to manage their building in a fire or an emergency situation. The FEF must always contain the following:

- all assumptions in the design of the fire safety systems such as fire load, any risk assessments or risk analysis;
- all assumptions in the design of the fire safety arrangements regarding the fire safety management of the building including emergency procedures;
- escape routes, escape strategy and muster points;
- details of all passive fire safety measures, for example, compartmentation, cavity barriers, fire doors, duct dampers and fire shutters;
- details of fire detector heads, smoke detectors, alarm call-points, fire safety signage, emergency lighting, dry or wet risers and other firefighting equipment, exterior facilities for fire and rescue services;
- details of all active fire safety measures such as sprinkler systems or smoke control systems;
- information about any elements of the fabric and services that may adversely affect the "general fire precautions" in a fire, for example, cladding;
- any other high-risk areas in the building, such as heating machinery;
- information on the requirements relating fire safety equipment, including operational details, manuals, software, routine testing, inspection and maintenance schedules; and
- provisions incorporated into the building to facilitate the evacuation of disabled and other potentially vulnerable people.

Specifically, the FEF must always contain a fire strategy design report setting out how the building satisfies Parts B1 to B5 of Schedule 1 of the Building Regulations. It should also include all the relevant technical specifications, product datasheets, operation and maintenance manuals and all inspection and commissioning records.

Full plans

"Full plans" are defined as a complete set of accurately recorded "as-built" drawings and a detailed specification of works, setting out how fire and structural safety risks will be managed with all amendments that have been made on site during the construction phase.

The construction control plan

The construction control plan must set out in detail how compliance with the Building Regulations will be achieved, including information on the change management system which states how changes to the design of a building will be controlled and recorded. Any major amendments which affect the safety of the building and/or the users of the building will need to be agreed with the Client, and the Principal Designer and be approved by the BSR.

The implementation of building information modelling

The draft Building Safety Bill states that the golden thread of building information will be held digitally to ensure that the original design intent, and any subsequent changes are recorded and can be used to support and inform building safety improvements. The Hackitt Review promotes Building Information Modelling (BIM) as a suitable vehicle for the creation and retention of the golden thread of information, reflecting the conclusion of the Government-commissioned report, "Modernise or Die", which views BIM the road map to collaboration and high efficiency for the UK construction industry. The report states that *"the importance of BIM adoption cannot be overestimated"*.[14] Over the last decade, BIM has developed into a relatively widely known digital tool in the construction industry, not least because in 2016 the Government issued a mandate to make all centrally procured Government contracts BIM Level 2 compliant.

BIM creates value throughout the lifecycle of an asset and is underpinned by the creation, collation and exchange of shared, three-dimensional (3D) models and intelligent data, such as technical specifications, and asset management information, which are attached to the models.[15] A typical construction project generates an enormous amount of data, which is often unstructured and poorly co-ordinated. The use of BIM as a digital platform, where all changes lead automatically to an updated model, coupled with an effective Common Data Environment (CDE) for hosting this information, provides a collaborative and co-ordinated environment which facilitates informed decision-making and leads, ultimately, to the reduction of risk.[16,17]

For new builds, the duty holders will use BIM to collect and record the golden thread information during the design and construction phases. Duty holders who use BIM effectively will be able to keep an accurate record of the "as-built" design of a building to satisfy the requirements of the BSR before the building is signed off for occupation and the golden thread of information is handed over to the AP.

The draft Building Safety Bill makes provision for the establishment of a "Mandatory Occurrence Reporting System" which requires that any structural and fire safety occurrence which may cause a significant risk to safety is reported to the BSR. The use of BIM would provide the appropriate framework that all duty holders must establish to facilitate mandatory occurrence reporting throughout the life cycle of a building.

The asset information model and the digital twin[18]

As a database, BIM's wide-ranging capability to store large amounts of intelligent data lends itself to a role in the design and construction phase of a building to produce the Project Information Model (PIM), but more productively, BIM can be used for the whole life management of an asset where the PIM informs the Asset Information Model (AIM) used in the operational phase of a building's life.

The characteristics and intended use of the AIM are detailed in PAS 1192-3, though this soon to be replaced by ISO 19650-3. The definition of the AIM states that it should include:

- information concerning the original brief, specification and design;
- a 3D object-based model or models of the environmental location of the asset. This could be a revised or "collapsed" version of the PIM;
- information about the ownership of the asset and data obtained from maintenance, surveys or other works carried out on the asset during its lifetime; and
- data obtained from monitoring the operation and condition of the asset.

The AIM as envisaged in PAS 1192-3 would seem a suitable vehicle for the "golden thread of building information" but it may be overtaken by the concept of the "Digital Twin". The Centre for Digital Built Britain defines a Digital Twin as *"a realistic digital representation of assets, processes or systems in the built or natural environment"*.[19] The distinguishing feature of a Digital Twin is its connection to the physical twin. The maturity spectrum for the development of Digital Twins set out in the Institution of Engineering and Technology report *Digital Twins for the Built Environment (2020)*[20] envisages a logarithmic increase in the scale of complexity from 0 to 5, starting with reality capture via scans, photography or even drawings, through to the autonomous operational maintenance of a building with complete self-governance and total oversight and transparency using the Digital Twin. Although this final stage is still in development, Stage 2 of the Digital Twin maturity spectrum is defined as the connection of a model to BIM Stage 2. It is reasonable to assume that, at this point, the AIM overlaps with the Digital Twin and the AIM will be a stage in the development of the Digital Twin.

The increased use of AIMs and the potential development of Digital Twins can certainly provide a platform to deliver the requirement for a golden thread of building information. It is, however, acknowledged that BIM in itself is not a "silver bullet", and there are barriers to the wholesale introduction of BIM envisaged by the Hackitt Review and in the draft Building Safety Bill.

Challenges to BIM implementation

There are, currently, barriers to the widespread adoption of BIM within the UK construction industry.[21] The UK construction industry is comprised of a wide range of actors of varying sizes and resources. The majority of construction firms are small or medium-sized enterprises with fewer than 250 employees. Their workforce will have limited technical capacity and may not possess the expertise required to use BIM collaboratively.[22] Site-based operatives would need to report on the changes to work specifications and drawings that occur on a regular basis during the currency of a project and may not have the technical competence to input information into a PIM. This may become a new role for a technically astute clerk of works appointed by the Client who is tasked with monitoring and recording all changes within the PIM.

When using BIM, an effective CDE can provide an optimum collaborative environment which is beneficial to all parties. BIM, however, is not a standardised "off-the shelf" technology,[23] and for each project, the parties involved will need to agree the scope of the data to be captured and in what file formats, particularly as some contractors and installers still transfer information using analogue PDFs and printed files.[24] Standardisation and harmonisation of data exchange are key issues for the inter-operability of BIM and the ability to capture and share consistent asset data that is standardised remains a challenge for the UK construction industry.[25]

It has been recognised that AIMs can have interface issues with existing client databases in that they don't "speak" to each other or readily accept data transfer between them.[26] There are two possible consequences in this situation. Either two databases operate for the same building, neither of which will be adequately updated, or one database will simply be ignored and become redundant, even though it may hold information vital to the building safety case or fire risk assessment.

The principal barrier to BIM adoption is that while BIM is widely known within the industry, it is not as widely understood.[27] BIM is an emergent technology as opposed to being static and widely used. It may be that the requirement to provide the "golden thread of building information" will prove to be a driver for the implementation of BIM, if only around the issues of building fire safety and structural safety to fulfil the regulatory requirements. The most significant factor in the successful implementation of BIM as the digital platform for the "golden thread" will be the competence of the actors who input information into the AIM and of the actors who access and use the data contained in the AIM.[28]

Product traceability

The Hackitt Review cites confusion over product labelling as a contributory factor to fire safety systems being compromised and recommends that the digital record will also be used to provide product traceability. The review also points out that the UK construction industry has been slow to adopt traceability and other quality assurance techniques which are in widespread use in other industrial and commercial sectors around the world.

Product traceability is currently used within the industry as it is sometimes a contractual requirement. Since 1 July 2013, manufacturers of construction products have been required to provide a declaration of performance and CE mark for products marketed and used within the EU. This means being able to trace a material or product, via recorded documentation, back to its source and verify every step of the manufacturing and distribution process. In most cases, products are physically labelled, using rip-proof or high-performance tags, attached to the product itself. However, the Hackitt Review recommends that traceability, with a clear, consistent labelling regime, should make use of readily available digital technologies, to fulfil the following objectives:

a   to provide confidence, assurance and the ability to check that the right product has been installed correctly, complies with the relevant standards and has been maintained in the appropriate way during the operation of a building;

b   to ensure the required product information is readily available so that action, such as product recall, can be taken in the event of non-compliance; and

c   to provide the ability to access all information about a product, offering potential for better quality control, improved procurement and support for continuous development and learning from experience.[29]

The Hackitt Review recommends the adoption of a traceability system but does not provide a formal definition of what traceability is or how it will be implemented. In the past, traceability has been defined by linking it to the need for accurate client requirements to be captured at the commencement of a project, with traceability subsequently established[30] to allow design decisions to be traced to implicit and explicit client requirements.[31] This definition has been extended to incorporate the ability to follow the information related to a product as it moves along the supply chain.[32]

In 2019, a study was developed at Northumbria University to propose a digital record for traceability of all built assets, new and existing. The study built on traceability systems research in other industries such as the food industry which has effective systems in place to maintain the highest levels of traceability for food safety and quality assurance. One of the main objectives of the Northumbria University research was to develop a definition of traceability specifically for the UK construction industry. The study built upon a definition of traceability applied in the food industry[33] and adapted it to produce the following definition for built asset traceability:

> The ability to record all required information relating to that which is under consideration, throughout its entire lifecycle, by means of recorded identifications.

The ability to ensure that all required information is readily available and in the correct format when needed will be central to the success of digital record-keeping. Information quality is as important as information traceability as the available information will drive any decision-making process.[34] It is self-evident that if the available information is incorrect, imprecise or simply not accessible, then this will negatively affect the process.[35]

As a result of a series of workshops, involving experienced practitioners from all areas and sectors of the UK construction industry supply chain, the following definition was produced to encapsulate Hackitt's requirements for a robust digital record of traceability:

> A digital record provides traceability through a secure, immutable and auditable electronic record of all required information, actions and decisions taken to assess and achieve compliance of a built asset with relevant standards and regulations at a point in time. It must record stakeholder and compliance requirements, design intent, procurement of materials/components and construction together with the testing, validation and verification processes undertaken, capturing their outcomes in order to provide a complete decision-trail. The record will include physical asset and performance data of all components and support traceability of provenance from raw material state through manufacturing, installation, maintenance and disposal, detailing who did what, when, why, how and to what specification. The digital record must be accurate, traceable, appropriately open, non-proprietary, searchable, and show clear delineations of risk ownership.

How is the record to be kept?

In recommending the use of a digital record, the Hackitt Review suggested learning from other sectors such as the aviation or automotive sectors, where traceability using digital technology is already embedded and Radio Frequency Identification (RFID) technology, quick response codes (QR) inkjet printing and nano particles are used to support traceability. Some of these technologies are already in use within the construction industry. RFID, a technology based on the exchange of information via electromagnetic signals has been in use for the last two decades[36] in a variety of applications, including the tracking and management of materials and tools.

Research has shown that emerging technologies, other than BIM, can support enhanced traceability. Distributed Ledger Technology (DLT) offers immutable, secure processing and recording of transactions across a distributed, decentralised peer-to-peer network. DLT can provide a potential storage mechanism for digital records and can support historical record-keeping which could provide an ideal ecosystem for traceability of materials.[37]

Smart contracts are, in essence, a digital contract which can automatically execute its terms when pre-defined conditions are met and can be used to increase efficiency and traceability.[38]

Through Blockchain, a Digital Twin can ensure traceability of the data it holds around operational strategies or preventative maintenance, etc. If part of the structure of a building fails unexpectedly, then through the Digital Twin, it would be relatively straightforward to ascertain which elements caused the failure, which organisation was responsible for its assembly and for its procurement.[39]

The British Standard that is being developed to manage the flow and presentation of information relevant to fire safety (*BS 8644 Digital Management of Fire Safety Information for Design, Construction, Handover and Emergency Response Code of Practice*) is "technology agnostic" and will provide protocols and guidance for those who wish to operate within a BIM framework and for those who do not. While the construction industry must embrace the digital era for fire safety information management, BS8644 will allow users of the standard to carry out a traditional information exchange using any digital format they choose.[40]

Summary

Dame Judith Hackitt was surprised to discover that documentation passed on to building owners by construction professionals lacks complete, accurate and up-to-date information on how a building has been constructed or refurbished. In other sectors, from cars to food production, it is accepted that every single part of a product can be traced and identified. Likewise, it is essential that the construction industry embraces the concept of the digital "golden thread" of information as an essential component in delivering the necessary improvements in building safety. Dame Hackitt has urged construction materials manufacturers to make greater efforts to digitise their processes as, currently, product testing, marketing, labelling and approval processes are flawed and unreliable.[41]

It has been suggested that it will take at least two years for the "golden thread" to become a standard part of the handover documentation for higher-risk buildings. Almost half of all client organisations and facilities management firms do not have the necessary software or the required technical capabilities to check that the information provided to them during the design and construction phases is accurate.[42] The current culture within the construction industry is perhaps the biggest challenge to overcome to ensure that the golden thread and traceability of products are implemented effectively. We need to make significant advances in education and learning to improve the understanding and acceptance of digital working within the construction industry. For the adoption of BIM to be successful across the whole of the industry, the negative perceptions around BIM and associated software need to change through learning and experience.[43] Only then can the digital "golden thread" of information be implemented correctly and deliver its purpose effectively.

Notes

1 Regulation 38 of the Building Regulations (2010) is a requirement to provide fire safety information to the 'Responsible Person' at the completion of a project, or where the building or extension is first occupied. However, in practice the regulation is not always fully complied with.
2 See page 57 of Hackitt, J Dame (2018) Building a Safer Future, Independent Review of Building Regulations and Fire Safety: Final Report.
3 Audit Commission (2001) "Acting on Facts". Audit Commission.
4 See page 13 of the Explanatory Notes to the Draft Building Safety Bill: Ministry of Housing, Communities and Local Government (2020) Draft Building Safety Bill: Explanatory Notes. Draft_Building_Safety_Bill_ PART_2.pdf (publishing.service.gov.uk) (Last Accessed 19/2/2021).
5 See page 110 of the Explanatory Notes to the Draft Building Safety Bill: Ministry of Housing, Communities and Local Government (2020) Draft Building Safety Bill: Explanatory Notes.
6 See page 35 of Hackitt, J Dame (2018) Building a Safer Future, Independent Review of Building Regulations and Fire Safety: Final Report.
7 See page 57 of Hackitt, J Dame (2018) Building a Safer Future, Independent Review of Building Regulations and Fire Safety: Final Report.
8 See page 102 of Hackitt, J Dame (2018) Building a Safer Future, Independent Review of Building Regulations and Fire Safety: Final Report.
9 See page 148 of Hackitt, J Dame (2018) Building a Safer Future, Independent Review of Building Regulations and Fire Safety: Final Report.
10 See page 102 of Hackitt, J Dame (2018) Building a Safer Future, Independent Review of Building Regulations and Fire Safety: Final Report.
11 See page 104 of Hackitt, J Dame (2018) Building a Safer Future, Independent Review of Building Regulations and Fire Safety: Final Report.
12 See page 56 of Hackitt, J Dame (2018) Building a Safer Future, Independent Review of Building Regulations and Fire Safety: Final Report.
13 See pages 132–133 Appendix D of Hackitt, J Dame (2018) Building a Safer Future, Independent Review of Building Regulations and Fire Safety: Final Report.
14 Farmer, M. (2016) Modernise or Die: The Farmer Review of the UK Construction Labour Model. Construction Leadership Council, p. 36.
15 Gov.UK (2012) Building Information Modelling. Industrial Strategy: Government and Industry in Partnership. Department of Business, Innovation & Skills.
16 Da Silva (2019) Using the Golden Thread to Reduce Risk and Improve Project Outcomes in Housing. BIM +, CIOB. Using the 'golden thread' to reduce risk and improve project outcomes in housing | BIM+ (bimplus. co.uk) (Accessed 14/2/2021).
17 BiM Regions (2016) The UK BIM Alliance – Cross Industry Alliance Emerging to Drive Leadership Around Implementation of BIM Level 2.
18 The section relies predominantly on Lewis, S (2020) Aims Assets and Threads. Joint Contracts Tribunal AIMs Assets and Threads – The Joint Contracts Tribunal (jctltd.co.uk) (Accessed 14/02/2012).
19 Centre for Digital Built Britain (2018) The Gemini Principles. Cambridge University, UK.

20 Institute of Engineering Technology (2020) Digital Twins for the Built Environment IET/Atkins. leaflet-digital-twins-for-the-built-environment-iet-atkins.pdf (theiet.org) (Accessed 14/02/2021).

21 Dainty, A, Leiringer, R, Fernie, S & Harty, C (2017). BIM and the Small Construction Firm: A Critical Perspective. *Building Research & Information* Vol 45(6) pp 696–709.

22 Spinardi, G (2020) Hackitt and the Golden Thread: Challenges to Implementation. Buildings & Cities. Hackitt and the Golden Thread: Challenges to Implementation – Commentaries (buildingsandcities.org) (Accessed 14/2/2021).

23 Spinardi, G (2020) Hackitt and the Golden Thread: Challenges to Implementation. Buildings & Cities. Hackitt and the Golden Thread: Challenges to Implementation – Commentaries (buildingsandcities.org) (Accessed 14/2/2021).

24 Cousins, Stephen (2019) Golden Key Could Unlock Hackitt's Golden Thread of BIM Data. BIM Golden key could unlock Hackitt's golden thread of BIM data | BIM+ (bimplus.co.uk) (Accessed 14/2/2021).

25 Cousins, Stephen (2019) Golden Key Could Unlock Hackitt's Golden Thread of BIM data. BIM Golden key could unlock Hackitt's golden thread of BIM data | BIM+ (bimplus.co.uk) (Accessed 14/2/2021).

26 Phillips, S & Foreman, T (2018) The Role of Building Information Modelling in Retrofitting Works within the UK Social Housing Sector. *Journal of Building Survey, Appraisal & Valuation* Vol 7(3) pp 229–245.

27 Phillips, S & Streatfield, C (2019) How can Digital Technology Improve Productivity in Retrofitting Works within the UK Social Housing Sector? ARCOM Conference 2 September 2019.

28 Spinardi, G (2020) Hackitt and the Golden Thread: Challenges to Implementation. Buildings & Cities.

29 Watson, R, Kaseem, M & Li, J (2019) Traceability for Built Assets: Proposed Framework for a Digital Record. Creative Construction Conference 29 June–2 July 2019, Budapest, Hungary.

30 Hao, Q, Shen, W, Neelamkavil, J & Thomas, JR (2008) Change Management in Construction Projects, in: Proc. CIB W78 25th Int. Conf. Inf. Technol. Improv. Management. Constr. Project. Through IT Adoption, Santiago, Chile, pp 387–396.

31 Kamara, JM, Anumba, CJ & Evbuomwan, NFO (2000) Establishing and Processing Client Requirements—A Key Aspect of Concurrent Engineering in Construction. *Engineering, Construction and Architectural Management* Vol 7 pp 15–28. https://doi.org/10.1108/eb021129

32 Katenbayeva, A, Glass, J, Anvuur, A & Ghumra, S (2016) Developing a Theoretical Framework of Traceability for Sustainability in the Construction Sector. Presented at the 12th Corporate Responsibility Research Conference, Istanbul, Turkey, 12–14 October.

33 Olsen, P & Borit, M (2018) The Components of a Food Traceability System. *Trends in Food Science & Technology* Vol 77 pp 143–149. https://doi.org/10.1016/j.tifs.2018.05.004

34 Li, J, Kassem M & Watson, R (2020) A Blockchain and Smart Contract-Based Framework to Increase Traceability of Built Assets. In Proc. 37th CIB W78 Information Technology for Construction Conference, Sao Paulo, Brazil.

35 Jylhä, T & Suvanto, ME (2015) Impacts of Poor Quality of Information in the Facility Management Field. *Facilities* Vol 33 pp 302–319. doi: 10.1108/F-07-2013-0057.
36 Valero, E, Adan, A & Cerrada, C (2015) Evolution of RFID Applications in Construction: A Literature Review.
37 Li, J, Kassem M & Watson, R (2020) A Blockchain and Smart Contract-Based Framework to Increase Traceability of Built Assets. In Proc. 37th CIB W78 Information Technology for Construction Conference, Sao Paulo, Brazil.
38 Penzes, B (2018) Blockchain Technology in the Construction Industry: Digital Transformation for High Productivity, Institution of Civil Engineers. Blockchain-technology-in-Construction-2018-12-17.pdf (ice.org.uk) (Accessed 14/02/2021).
39 Penzes, B (2018) Blockchain Technology in the Construction Industry: Digital Transformation for High Productivity, Institution of Civil Engineers.
40 Chevin, D (2020) The New Fire Safety Digital Framework Explained. BIM+ The new fire safety digital framework explained | BIM+ (bimplus.co.uk).
41 BIM + (2020) Hackitt Urges Materials Producers to Digitise Faster. Hackitt urges materials producers to digitise faster | BIM+ (bimplus.co.uk) (Accessed 19/2/2021).
42 CIOB & i3PT (2020) The Golden Thread: Understanding the Capability and Capacity of the UK Built Environment to Deliver and Retain Digital Information. Golden-Thread-Review.pdf.
43 Phillips, S & Foreman, T (2018) The Role of Building Information Modelling in Retrofitting Works within the UK Social Housing Sector. *Journal of Building Survey, Appraisal & Valuation* Vol 7(3) pp 229–245.

6 Four years on

Background[1]

Clearly, progress has been made towards establishing the new regime for building safety. The draft Building Safety Bill has been published, though the detail will need to be set out in secondary legislation. Dame Judith Hackitt has been made chair of the Transition Board, tasked with establishing the new BSR, and Peter Baker has been announced as the first Chief Inspector of Buildings to lead the new BSR.[2] However, progress has been slow and difficult in other areas especially with respect to the remediation works to HRRBs.

In the wake of the Grenfell Tower fire, the government set up an independent expert advisory panel (the expert panel) to provide advice and make recommendations to the MHCLG on immediate building safety measures that should be implemented to identify HRRBs of 18 m or more in order. The 18 m threshold is stated in Approved Document B to the Building Regulations as the point at which additional fire safety provisions are required.

The expert panel combines a wealth of experience and knowledge around fire prevention, building safety and testing processes with a remit that allows them to engage additional technical expertise if necessary. The expert panel includes a former London Fire Commissioner, the Chair of the National Fire Chiefs Council and the Chief Executive of the Building Research Establishment.

The original terms of reference[3] for the expert panel included the provision of advice on fire safety issues in relation to: (a) Aluminium Composite Material (ACM) cladding systems including polyethylene fillers which was the material used at Grenfell Tower, (b) fire doors and other building safety products and (c) assisting the government in implementing the Hackitt Review. The expert panel was tasked to consider whether any changes or clarifications to existing regulations

DOI: 10.1201/9781003092803-6

were necessary. Similarly the panel also considered whether current processes for assessing building safety were adequate and whether any amendments were necessary. Specifically excluded from the remit of the expert panel was the determination of government policy, nor could they cut across the public inquiry and police investigations into the Grenfell Tower fire.

On advice received from the expert panel, the government amended Regulation 7 of the Building Regulations 2010 to effectively ban the use of combustible materials in the external walls and balconies, of new buildings over 18 m in height with effect from 21 December 2018. Only materials classified A1, A2-s1, d0 as non-combustible are permissible and no other combustible materials may be used.[4] The government also directed that ACM external wall systems should be removed from HRRBs and provided support for local authorities to undertake emergency works to remove and replace ACM cladding.

Advice notes

In addition to the amended legislation addressing the issues around ACM cladding, the MHCLG has also published a series of fire safety advice notes detailing what owners of HRRBs should do to satisfy their obligation to take reasonable fire precautions to ensure the building is safe for residents as required by the Regulatory Reform (Fire Safety) Order 2005. The MHCLG issued 22 advice notes in total, and in January 2020, these were consolidated into a single document entitled "Advice for Building Owners of Multi-storey, Multi-occupied Residential Buildings" ("the consolidated advice document").[5] In order to understand the full implication of this document for the valuation of HRRBs, it is necessary to explore the effect that Advice Note 14 (AN14)[6] had on the market prior to the publication of the consolidated advice document.

In December 2018, the MHCLG issued AN14 which covered the fire safety of external wall systems on those buildings above 18 m in height which do not contain ACM within the cladding system. AN14 confirmed the government's endorsement of the recommendations of Dame Judith Hackitt in her Independent Review of Building Regulations and Fire Safety.[7] The scope of AN14 encompassed all HRRBs that have any type of external cladding system, other than ACM, which is potentially combustible including Metal Composite Cladding (MCM) faced with other materials such as zinc, copper, and stainless steel, High Pressure Laminates (HPL) and rendered insulation systems.

AN 14 also highlighted the fact that a Responsible Person under the Regulatory Reform (Fire Safety) Order 2005 (RRO 2005) has a duty to consider "general fire precautions" requiring them to reduce the risk of fire and the spread of fire within the building they are responsible for. The MHCLG implied in the advice note that the scope of this duty has now been extended so that it includes the Responsible Person (or their agent) assessing and reviewing the construction of the external wall systems as part of a Fire Risk Assessment. This includes, not only the type of external panels that are used, but also insulation, cavity barriers, fire stopping and other elements of the external wall detailing.

One of the main problems with AN 14 was that it contained an extremely narrow definition of whether an external wall system is safe or not. With regard to existing buildings, an external wall system can only be safe if it is comprised of materials of specified limited combustibility or it has successfully passed the BS8414 test and achieved BR 135 classification as set out in the Building Research Establishment (BRE) publication "Fire Performance of external thermal insultation for walls of multi-storey buildings".[8] This issue is exacerbated by the fact that even if a building owner is able to demonstrate that their system complies with the BS8414 test they must also show that that the system is demonstrably safe, that is, that the construction of the system on the building unequivocally matches the system that was tested and that it has been correctly installed and adequately maintained.[9] The definition of "safe" does not include any reference to fire engineered solutions that may have been implemented within an HRRB and so assessments cannot take into account the amount of any material used, whether sprinklers have been installed or even if a waking watch system has been implemented.

Manifestly the external wall systems of the majority of HRRBs will not have been subject to the BS8414 test. On the contrary, it would be reasonable to assume that designers would have sought compliance by applying the guidance provided within Approved Document B of the Building Regulations. We are now aware the guidance and recommendations stated in the Approved Document have often been applied incorrectly.[10] As a consequence, it is also reasonable to assume that most HRRB owners will now be unable to demonstrate that the external wall system to their building is safe and they will need to commission surveys or inspections to ascertain whether or not the materials used in the external wall system have been adequately installed and maintained and, where appropriate, determine whether the construction of the external wall system complies with the BS8414 test.

The most significant consequence of AN14 is that it has, in effect, created a retrospective obligation on all HRRB owners not only to undertake inspections to ensure that the construction of the external wall systems is safe but more importantly, to undertake remedial works to any external wall systems that is not safe. In addition to this, the owners will also need to implement mitigation measures until the wall can be deemed to be "safe".

The publication of the consolidated advice document in January 2020 showed that the expert panel had not simply amalgamated all the previously issued advice notes into one document. It has also extended the remit of the advice so that owners of all multi-occupied residential premises of any height must consider and mitigate the risks of any external wall system and fire doors when discharging their duty under the RRO 2005. The remit was extended because the expert panel had serious concerns that Requirement B4 of Schedule 1 of the Building Regulations was not considered in full detail. The report states:

> Requirement B4 is clear and requires that "the external walls of the building shall adequately resist the spread of fire over the walls and from one building to another, having regard to the height, use and location of the building." The need to assess and manage the risk of external fire spread applies to buildings of any height.[11]

In effect, the definition of an HRRB has been extended to include all multi-occupied residential buildings, with cladding or without cladding and of any height. The expert panel also advised that building owners should not wait for the regulatory system to be reformed but should actively be checking the construction of their buildings to ensure the safety of their residents.

Valuation issues

Though the MHCLG have stated that the consolidated advice document is not a statutory instrument nor a compliance document and is only intended to assist building owners to take action should they have any concerns over the safety of their external wall systems it has had a significant, albeit unintended, effect on the valuation market. This has occurred because building owners are required to provide technical information about the construction of their buildings prior to a lending decision being made. However, many freeholders simply do not hold the information that is required not least because the owners must provide absolute assurances that there is no combustible material contained within the tall building.

Many building owners will be property organisations that own multiple HRRBs and may not be able to locate all the appropriate records including as-built information necessary to determine what materials make up their external walling systems. Even if the data can be located, it may be inaccurate or unclear on the classification of materials used. For instance, materials of limited combustibility are defined by reference to the Building Regulations that were in force at the time of construction and expert advice may be needed to be sought to determine the exact nature of the materials used. The problem is exacerbated where a building contains material which is not of limited combustibility, as there is a much greater likelihood that the HRRB will not be considered safe.

As a result of the implementation of AN14 and owners of HRRBs checking compliance of their properties with the requirements of the Building Regulations, the industry has become aware of a serious problem with both the design and construction of external wall systems. The defects are not confined to the use of combustible external cladding materials but include the forms of insulation used and the installation of defective cavity barriers.[12] Consequently, this issue has caused valuers and lenders difficulties in producing a valuation for an HRRB and the individual flats therein. If the necessary technical information cannot be provided, then lending institutions have instructed valuers to put a £nil value against the property ensuring that it cannot pass through any credit approval process. A £nil valuation does not mean a flat is worthless, but it prevents a mortgage advance being made against a property at least until more information from an intrusive inspection by competent professionals becomes available and an accurate valuation can be produced.[13]

However, carrying out an inspection of HRRB to demonstrate compliance with the consolidated document can be a complex process, and the Royal Institution of Chartered Surveyors (RICS) has stated that cladding and fire safety maybe the biggest technical challenge the profession has faced in several decades. A typical inspection may include the input of a chartered fire engineer and from a specialist cladding engineer with a contractor in attendance. The services of a testing laboratory will be required to determine the materials used in the cladding system. It is crucial to understand whether the existing cavity barriers are adequately fixed so samples from multiple points within the cladding system may need to be taken to provide a comprehensive picture of how the barriers have been fitted and whether they are fit for purpose. Due to these issues, Professional Indemnity Insurance (PII) providers are reconsidering policy coverage in order to reduce their exposure to fire safety claims due to external cladding systems.[14]

The unintended consequence of these various factors is that sales of new flats and the re-mortgaging of existing flats including shared ownership staircasing transactions are placed in abeyance. Valuers and lending institutions are insisting on assurances from competent professionals that buildings comply with the new guidance and the steps set out in the consolidated advice document have been successfully carried out and confirm the building is either safe or does not present a significant fire risk. It has been estimated that between one[15] and three million[16] flats could be impacted by the problem with between four and five million residents affected.

The external fire review process

In order to address this blockage within the mortgage market, the Royal Institution of Chartered Surveyors (RICS) led a cross-industry working group to determine best practice in the reporting and valuation of tall buildings. On 19 December 2019, the RICS together with the Building Societies Association (BSA) and UK Finance agreed a new External Wall Fire Review, which is a standardised process for the valuation of HRRBs with actual or potential combustible materials to external wall systems and balconies. The review process (see Figure 6) requires a fire safety assessment to be undertaken by a qualified, competent construction professional. Only one assessment is required per building and the assessment will be valid for a five-year period. It is envisaged that the new industry-wide process can be used by valuers, lending institutions, freeholders, construction professionals and fire safety experts and will help people buy or sell homes and re-mortgage flats in HRRBs.[17]

The EWS form can only be completed by competent, chartered professionals with the requisite experience of fire design and safety within buildings and cannot be completed by the valuer even though they may have the necessary fire expertise. The EWS provides two options for recording the findings of an HRRB inspection. Option A is completed where the primary materials (i.e. the cladding system, forms of insulation system, etc.) do not support combustion. Under this option a competent professional will certify that the HRRB contains materials that do not support combustion or while it does contain some materials that support combustion, they are limited to attachments or additions to the external wall (e.g. balconies) and do not require high value remedial works. Remedial works will normally be subject to guidance supplied by the relevant lending institution and combustible materials to, say, balconies could be reflected either in the valuation itself or by the retention of a set amount. Where lender guidance is not provided or does not reflect the impact on the valuation then it would be reasonable to assume that necessary remediation costs would be detailed and reflected in the property valuation.[18]

The second option, Option B, is invoked where the primary materials in the external wall system are known to be combustible. In this case, a Chartered Fire Engineer (or equivalent) will need to undertake a risk assessment to provide an opinion on whether the fire risk is low, or whether remediation works will be required. In order to make this determination the fire engineer will need to assess whether or not the external wall system of the HRRB complies with the functional requirements of the Building Regulations, that is, does it resist the spread of fire and smoke so as to inhibit the spread of fie within the building? Will the unseen spread of fire and smoke be inhibited within confined spaces, and does it resist the spread of fire over the walls with regard to the height, position and use of the building?

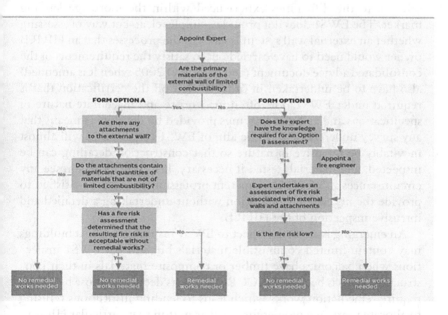

Figure 6 The EWS Process (Credit: Inside Housing[19]).

Discussion: the problems with EWS1

The core aspiration of the EWS process is to move away from the binary process invoked by AN14 and reinforced within the consolidated advice document which advises that all external walls on HRRBs must be checked and if the materials are combustible then they must pass the BS8414 test. The EWS process is meant to be a more nuanced one, in so far as the assessment of risk in respect of the fire safety of a building has been delegated to a competent chartered professional. Their role is either to identify HRRBs that need remediation work or to give assurance to lenders when buildings are deemed to be safe with respect to fire. The crucial part of this process is the assessment of risk as this was initially believed to be the factor that would start to unclog the mortgage market. The expert must assess whether only materials of limited combustibility are contained within the external wall system and/or attachments such as balconies. If this is the case then the building can be deemed safe for the purpose of the mortgage companies. Similarly, if the materials are found to be combustible but are deemed to be low risk, then the building can still be deemed to be safe for lending purposes. Both outcomes will restore fluidity to the mortgage market.

However, at the present time, it is clear that the EWS1 is not the solution to the difficulties experienced within the mortgage lending market. The EWS1 does not provide a simple, clear-cut way of assessing whether an external wall system is safe. All the processes that an HRRB owner would need to have carried out to satisfy the requirements of the consolidated advice document (and the RRO 2005 when it is amended) also have to be undertaken in order to sign off the certification that is required under EWS1. The often inadequate and inaccurate nature of specifications and as-built drawings provided by developers means that any survey undertaken with the aim of EWS1 certification will almost inevitably be intrusive in nature so that construction detailing can be inspected, and materials tested if necessary. It is difficult to foresee any circumstances in which a competent professional would be satisfied to provide the necessary certification without undertaking a detailed and intrusive inspection of the HRRB.

An emerging issue with respect to EWS1 inspections is that buildings may contain limited combustible materials but still fail EWS1 inspections when balconies have timber or composite materials in their construction. This is because the CCP/inspectors deem that those buildings require remediation works which leads to lending institutions refusing to allocate mortgage monies for any flats within that particular HRRB. Anecdotal evidence is suggesting that inspectors are recommending remediation works on most EWS1 forms. This may reflect the number of buildings that have combustible materials in some form within their external wall system and, perhaps, a reluctance on the part of inspectors to deem the presence of any combustible material to represent a low risk. This issue is compounded by the time it takes for the recommended remediation work to be undertaken which may mean that, in the worst case, scenario it could be some years before flats affected can be sold or mortgaged.[20]

The pragmatic difficulties involved in undertaking an EWS1 inspection are not the only issue delaying the external fire review process. The situation has been exacerbated by the low number of qualified inspectors available against the huge number of buildings that need to be surveyed. It has been estimated that there are fewer than 300 chartered fire engineers who can undertake an EWS1 inspection which has inevitably led to long delays in arranging inspections. It can reasonably be foreseen that this pool of engineers may become even smaller as concerns around liability issues will lead to them either refusing to carry out EWS1 instructions or being unable to obtain PII cover for this work. Post Grenfell, PII insurers have become increasingly cautious about providing cover for fire safety inspections due to the potential financial liability should a similar tragedy occur again. It is difficult to foresee any situation where an inspector would undertake an EWS1 inspection without having the benefit of PII cover. This cautious approach is not limited to insurers but extends to banks and, despite UK Finance backing the process, some lending institutions are disinclined to agree mortgages even after a signed and completed EWS1 form has been provided. The fact that not all banks and lending institutions accept EWS1 forms has added confusion to an already disordered situation.

The final factor that may potentially delay the external fire review process is the cost of the inspection itself, which can be between £10,000 and £50,000.[21] This means leaseholders in HRRBs may have to contribute substantial sums of money via their service charges simply to enable the survey process to commence, with the prospect that they may need to pay even more monies should remediation works be required. Reports of fraudulent inspections have added further complications to an already challenging scenario.[22] The EWS1 process was developed with the best of intentions to assist in resolving the difficult situation that has arisen. Whilst there is no doubt that the EWS1 form has been accepted by some lending institutions and sales on some flats are proceeding, it is clear that the external fire review system has not had the intended effect and the related property market is still not functioning effectively.

Summary

Four years on from the Grenfell Tower fire, no effective pragmatic approach has been formulated to address the external wall system issues that are affecting HRRBs all around the country. Initially, after Grenfell, the government focused on risk issues surrounding the installation of ACM cladding on buildings over 18 m in height and they moved quickly and correctly to form an expert panel to advise on the testing and replacement of these cladding materials. However, since then, the issuing of a number of advice notes including the widely criticised AN14, has turned the focus to multi-occupied residential buildings that may contain any combustible material, irrespective of height, and scrutiny applies not only to the external wall system but to attachments to the building such as balconies. The result is that the scope of buildings that need to be investigated for risk to fire safety has become so wide that the industry is unable to cope with the number of inspections that need to be undertaken, and the extent of the consequent remediation works required to ensure that HRRBs are deemed safe. The consequence of this is that, despite all the measures undertaken to date, residents are still living in HRRBs that are unsafe and, particularly in the private sector, it seems unlikely that the necessary remediation works will be carried out to their homes in the foreseeable future.

The advice notes have also had the unforeseen effect of blocking the market for the sale and re-mortgaging of flats. The EWS1 form has not had its intended effect of unlocking the market and, in retrospect the development of the external fire review process should perhaps have included initial input from representatives of the insurance industry. In its existing form, the process is virtually unworkable, not least because some insurers are loathe to provide cover for the inspections because the EWS1 form is considered to be a matter of concern to insurers across the market. The resolution to this complex situation will not be easy but easing the crisis and reducing the stress of concerned residents could be made by adjustments to the existing process and it would seem imperative that the insurance industry is included in these decisions. A relaxation of the rules on the professionals who are empowered to sign off option B in the EWS1 form could be considered. For example, in addition to Chartered Fire Engineers both Chartered Building Surveyors and Chartered Building Engineers could reasonably and safely carry out this function. Further risk analysis should be done to define which buildings should fall within the scope of the consolidated advice note and more explicit guidance to determine the correct prioritisation of HRRBs should be formulated.

In November 2020, the RICS announced a new eight-week training programme for Chartered Building and Building Control surveyors to qualify them to carry out external wall assessments on low to medium risk residential buildings, in an attempt to speed up the EWS1 process. Funding of £750,000 was provided by the Government to support the training scheme. However, for buildings that are over 18 m or those that require specialist testing, a qualified fire safety engineer is still required to carry out the EWS 1 assessments.[23] To underpin the creation of additional surveyors to sign off the EWS1 form, the Government has created a new professional indemnity insurance scheme to provide surveyors with the necessary professional support to sign off the forms. This, combined with the other initiatives, should enable some of the leaseholders caught in the EWS1 trap to sell or re-mortgage their flats.[24]

Government funding to assist residents living in HRRBs has also been increased, through the establishment of a £30 million "Waking Watch Relief Fund". This fund is available to finance the installation of common fire alarms in buildings where leaseholders have been forced to employ "waking watch" schemes so that the need for these schemes becomes redundant.[25] In February 2021, the Government also announced that it had set aside £3.5 billion to fully fund the replacement of unsafe cladding for all leaseholders in residential buildings that are 18 m and over in height, ensuring funding is targeted at the highest risk buildings. These costs will be paid for by the introduction of a "Gateway 2 developer" levy which will be applied when developers seek permission to develop certain types of high-rise buildings in England. Additionally, there are plans to introduce a new tax to be applied to the development sector, which is anticipated to raise £2 billion pounds to help pay for cladding remediation costs.[26]

Notes

1 The contents of this chapter are predominantly taken from a peer reviewed academic paper written by Steve Phillips with assistance from Jim Martin and Ross Hailey of Martin Arnold Chartered Surveyors entitled '*The Valuation of High-Risk Residential Buildings and the Role of EWS1*' which has been published by *The Journal of Building Survey, Appraisal & Valuation*.

2 Health & Safety Executive (2021) HSE Announces New Chief Inspector of Buildings. HSE 16 February 2021. HSE announces new Chief Inspector of Buildings – HSE Media Centre (Accessed 20/2/2021).

3 Ministry of Housing, Communities and Local Government (2018) *Independent Expert Advisory Panel – Updated Terms of Reference*. MHCLG/BSP/TOR/180918.

4   Building (Amendment) Regulations 2018: SI 2018/1230.

5   Ministry of Housing, Communities and Local Government (2020) *Advice for Building Owners of Multi-storied, Multi-occupied Residential Buildings.* MHCLG.

6   Department for Communities and Local Government (2018) *Advice Note 14: Advice for Building Owners on External Wall Systems that do not incorporate Aluminium Composite Material.* DCLG/BSP/Advice Note/ 14/171018.

7   Devonshires (2018) Beyond ACM: New Government Guidance on Fire Safety of External Wall Systems. https://www.devonshires.com/ publications/beyond-acm-new-government-guidance-on-fire-safety-of-external-wall-systems/ (Accessed 8/9/2020).

8   Building Research Establishment (2013) *Fire Performance of External Thermal Insultation for Walls of Multi-storey Buildings.* BRE, UK.

9   Building Research Establishment (2020) BRE Global – BRE 135 Classified External Cladding Systems. https://www.bre.co.uk/regulatory-testing (Accessed 3/9/2020).

10 Ministry of Housing, Communities and Local Government (2020) *Advice for Building Owners of Multi-storied, Multi-occupied Residential Buildings.* MHCLG.

11 Ministry of Housing, Communities and Local Government (2020) *Advice for Building Owners of Multi-storied, Multi-occupied Residential Buildings* (p 1). MHCLG.

12 Devonshires (2020) EWS 1: External Wall Fire Review and What It Means for Tall Building Owners. https://www.devonshires.com/publications/ ews1-external-wall-fire-review-and-what-it-means-for-tall-building-owners/ (Accessed 8/9/2020).

13 ARMA (2019) ARMA Update on Advice Note 14. ARMA https://arma. org.uk/downloader/tyq/ARMA_update_on_Government_Advice_ Note_14.pdf (Accessed 8/9/2020).

14 Royal Institution of Chartered Surveyors (2019) New Industry-wide Process Agreed for the Valuation of High-Rise Buildings. https://www. rics.org/uk/news-insight/latest-news/fire-safety/new-industry-wide-process-agreed-for-valuation-of-high-rise-buildings/ (Accessed 7/9/2020).

15 Wilson, R (2019) Industry Group Forms to Solve Advice Note 14 Issues. Social Housing December 2019. https://www.socialhousing.co.uk/news/ news/industry-group-forms-to-solve-advice-note-14-issues-64432

16 Dilworth, M (2020) The Red Tape Nightmare Stopping Millions Selling their Flats. This is Money. https://www.thisismoney.co.uk/money/ markets/article-8616399/Red-tape-nightmare-stops-millions-selling-homes.html (Accessed 12/9/2020).

17 Royal Institution of Chartered Surveyors (2019) New Industry-Wide Process Agreed for the Valuation of High-Rise Buildings. https://www. rics.org/uk/news-insight/latest-news/fire-safety/new-industry-wide-process-agreed-for-valuation-of-high-rise-buildings/ (Accessed 7/9/2020).

18 Royal Institution of Chartered Surveyors (2019) How Can Professionals Ensure that Cladding Is Actually Safe? https://www.rics.org/uk/news-insight/ latest-news/news-opinion/beneath-the-surface-how-can-professionals-ensure-cladding-is-actually-safe/ (Accessed 9/9/2020).

19 Inside Housing (2020) *Why the Attempt to Fix the Cladding and Mortgage Crisis Is Not Working.* Inside Housing. 20/2/20. https://www.insidehousing. co.uk/insight/insight/ews-why-the-attempt-to-fix-the-cladding-and-mortgage-crisis-is-not-working-65112 (Accessed 29/10/2020).

20 Simpson, J (2020) *Why the Attempt to Fix the Cladding and Mortgage Crisis Is Not Working.* Inside Housing. 20/2/20.

21 Oliver Fisher Solicitors (2020) What Is the EWS1 Form? How much does it cost, how long does it take, and why do I need it? https://oliverfisher. co.uk/what-is-the-ews1-form-how-much-does-it-cost-how-long-does-it-take-and-why-do-i-need-it/ (Accessed 11/9/2020).

22 Merryweather, L (2020) Scammers Hijack EWS1 Process with Fake Cladding Inspection Forms. Which 26 August 2020. https://www. which.co.uk/news/2020/08/scammers-hijack-ews1-process-with-fake-cladding-inspection-forms/ (Accessed 11/9/2020).

23 Royal Institution of Chartered Surveyors (2020) RICS Launches New Training Programme to Build Fire Safety Capacity. RICS 20 November 2020. RICS launches new training programme to build fire safety capacity (Accessed 20/2/2021).

24 Thomas, O (2021) Government Launching PII Scheme to Support EWS1 Process. Mortgage Solutions 10 February 2021. Government launching PII scheme to support EWS1 process – Mortgage Solutions (Accessed 20/1/2021).

25 MHCLG (2020) New £30 million Waking Watch Relief Fund announced. Gov.UK 17 December 2020 New £30 million Waking Watch Relief Fund announced – GOV.UK (www.gov.uk) (Accessed 20/2/2021).

26 MHCLG (2021) Government to Bring an End to Unsafe Cladding with Multi-billion Pound Intervention. Gov.UK 10 February 2021. Government to bring an end to unsafe cladding with multi-billion pound intervention – GOV.UK (www.gov.uk) (Accessed 20/2/2021).

Glossary of terms[1]

Term	Notes
Accountable Person	The duty holder during a building's occupation.
Approved Inspector	Previous system name for a private sector building control body, now called a registered building control approver.
Approved Document B	Guidance on ways to comply with the fire safety requirements in Part B of Schedule 1 to the Building Regulations 2010.
Assessment in lieu of test	An assessment carried out in lieu of a physical test. The term is particularly associated with cladding systems and is also referred to as a "desktop study".
Building Advisory Committee	A new expert advisory committee set up by the Building Safety Regulator to provide advice and information to the Building Safety Regulator in relation to its functions.
Building Assurance Certificate	A certificate that an Accountable Person must apply for and the Building Safety Regulator will provide if it is satisfied that the Accountable Person is complying with meeting its statutory obligations.
Building control authority	A generic name used for local authorities and the Building Safety Regulator in situations where either may be responsible for Building Act matters or checking compliance with Building Regulations' requirements.
Building Regulations Advisory Committee (BRAC)	Advisory committee established under (former) section 14 Building Act, 1984, for the purpose of advising the Secretary of State on the exercise of the Secretary of State's power to make building regulations, and on other subjects connected with building regulations. This committee is to be abolished and be replaced by the Building Advisory Committee.

(Continued)

Term	Notes
Building Safety Manager	Named by the Accountable Person, the Building Safety Manager will support the Accountable Person by carrying out the day-to-day functions of ensuring that the building is safely managed and promote openness, trust and collaboration with residents.
Building Safety Regulator	The Building Safety Regulator will be set up within the Health and Safety Executive, and make buildings safer through the implementation and enforcement of the new more stringent regulatory regime for buildings in scope, stronger oversight of the safety and performance of all buildings, and assisting and encouraging competence among the built environment industry, and registered building inspectors.
Building safety risks	A risk to the safety of persons in or about a building arising from the occurrence of fire, structural failure.
Buildings in scope	Refers to those buildings (described in the Bill as "higher-risk buildings") which will be within the scope of the new regulatory regime provided for in Part 3 or 4 of the Bill
Committee on Industry Competence	A new industry-led, expert committee set up by the Building Safety Regulator to facilitate improvement in the competence of the built environment sector.
Compartmentation	Construction provided to prevent the spread of fire to or from another part of the same building or an adjoining building. For example, compartment walls and floors with a rated period of fire resistance are provided to separate individual flats.
Construction control plan	One of the core information products that duty holders must produce during the design and construction phase – it describes how building safety and Building Regulations compliance will be maintained during the construction phase and how any changes to the Full Plans signed off at Gateway Point 2 will be controlled and recorded.

Term	*Notes*
Duty holders	The key roles (whether fulfilled by individuals or organisations) that are assigned specific responsibilities at particular phases of the building life cycle.
Fire and Emergency File	One of the core information products that duty holders must produce during the design and construction phase and it must be handed over to the building owner on occupation. This file will contain specified information and will help the building owner to better understand how to effectively manage their building in respect to fire/emergency situation.
Fire risk assessment	A fire risk assessment is an organised and methodical look at the premises, the activities carried on there and the likelihood that a fire could start and cause harm to those in and around the premises. Under the Fire Safety Order, the aims of the fire risk assessments are to identify the fire hazards, to reduce the risk of those hazards causing harm to as low as reasonably practicable and to decide what physical fire precautions and management arrangements are necessary to ensure the safety of people in the premises if a fire does start.
Gateway one, two and three	Three key stages in the building development where the duty holder must demonstrate that they are managing building safety risks appropriately before they are permitted by the regulator to continue to the next stage of development.
Golden thread of information	Fire and structural safety building information held digitally to specific standards. These standards will include requirements around robust information management and keeping the information up to date. The golden thread will ensure that those responsible for the building have the required information to manage building safety during throughout the life cycle of the building.
Higher-risk buildings	The technical term for buildings in scope of the new more stringent regulatory regime, as defined in the Bill.

(Continued)

Term	Notes
Higher-Risk Residential Building (HRRB)	Multi-occupancy higher risk residential buildings (that are ten storeys or more in height). They are the primary focus of the new regulatory framework set out in the Hackitt Review.
The Housing Ombudsman	The Housing Ombudsman provides redress for social housing residents. Membership of the Scheme is compulsory for social landlords (primarily housing associations who are or have been registered with the social housing regulator) and local authority landlords.
Mandatory Occurrence Reporting System	The Mandatory Occurrence Reporting System will require duty holders across all stages of the building life cycle to report fire and structural safety occurrences to the Building Safety Regulator which could pose a significant risk to life.
New Homes Ombudsman	An independent third party to provide alternative dispute resolution service between developers and purchasers of new build homes, and to remedy complaints.
Outcomes-based system	The system defines the outcomes or performance level to be achieved not the way those outcomes must be met.
Prescriptive system	The system defines the prescribed criteria to be met not the outcome to be achieved. Meeting a desired outcome or performance level is presumed if the prescribed criteria are met.
Principal Contractor	Under the Construction (Design and Management) Regulations 2015, a principal contractor is a contractor appointed by the client to be in control of the construction phase of the project when there is more than one contractor working on the project.
Principal Designer	Under the Construction (Design and Management) Regulations 2015, a principal designer is a designer appointed by the client to be in control of the pre-construction phase of the project, when there is more than one designer working on the project. The principal designer can be an organisation or an individual.

Term	Notes
Registered building control approver(s)	Formerly known as an Approved Inspector or a building control body under the old regulatory system. Refers to private sector firms doing building control work.
Registered building inspector	Refers to individual inspectors that are registered by the Building Safety Regulator.
Residents' Engagement Strategy	This is the means by which those living in buildings covered by the new regulatory regime can get involved in the decision-making process in relation to the safety of their homes. It will set out the approach and the activities that the Accountable Person will undertake to deliver these opportunities for all residents to participate.
Residents' panel	A statutory committee to be set up by the Building Safety Regulator. The residents' panel will be made up of residents and representatives/advocates of residents and advise the Building Safety Regulator on strategy, policy, systems and guidance which will be of particular interest to residents of higher-risk buildings.
Responsible Person	Under the Regulatory Reform (Fire Safety) Order 2005, a responsible person for a workplace is the employer or, in premises which are not a workplace, the person who has control of the premises in connection with carrying on of a trade, business or other undertaking (whether for profit or not) or the owner.
Safety Case Report	A structured argument, supported by a body of evidence that provides a compelling, comprehensible, evidenced and valid case as to how the Accountable Person is proactively managing fire and structural risks in order to prevent a major incident and limit the consequences to people in and around the building.

Note

1 The glossary is based upon the glossaries that are contained within both the Explanatory Notes to the Draft Building Safety Bill and the Hackitt Independent Review: Final Report.

Index

Note: **Bold** page numbers refer to tables; *italic* page numbers refer to figures and page numbers followed by "n" denote endnotes.

Printed in the United States
by Baker & Taylor Publisher Services